THE ALPHABET AND THE ALGORITHM

Writing **Architecture** series

A project of the Anyone Corporation; Cynthia Davidson, editor

Earth Moves: The Furnishing of Territories
Bernard Cache, 1995

Architecture as Metaphor: Language, Number, Money
Kojin Karatani, 1995

Differences: Topographies of Contemporary Architecture
Ignasi de Solà-Morales, 1996

Constructions
John Rajchman, 1997

Such Places as Memory
John Hejduk, 1998

Welcome to The Hotel Architecture
Roger Connah, 1998

Fire and Memory: On Architecture and Energy
Luis Fernández-Galiano, 2000

A Landscape of Events
Paul Virilio, 2000

Architecture from the Outside: Essays on Virtual and Real Space
Elizabeth Grosz, 2001

Public Intimacy: Architecture and the Visul Arts
Giuliana Bruno, 2007

Strange Details
Michael Cadwell, 2007

Histories of the Immediate Present: Inventing Architectural Modernism
Anthony Vidler, 2008

Drawing for Architecture
Leon Krier, 2009

Architecture's Desire: Reading the Late Avant-Garde
K. Michael Hays, 2010

The Possibility of an Absolute Architecture
Pier Vittorio Aureli, 2011

The Alphabet and the Algorithm
Mario Carpo, 2011

THE ALPHABET AND THE ALGORITHM

MARIO CARPO

THE MIT PRESS

CAMBRIDGE, MASSACHUSETTS

LONDON, ENGLAND

MIT Press books may be purchased at special quantity discounts for business or sales promotional use. For information, please email special_sales@mitpress.mit.edu or write to Special Sales Department, The MIT Press, 55 Hayward Street, Cambridge, MA 02142.

This book was set in Filosofia and Trade Gothic by The MIT Press. Printed and bound in the United States of America.

Library of Congress Cataloging-in-Publication Data

Carpo, Mario.
The alphabet and the algorithm / Mario Carpo.
p. cm. — (Writing architecture)
Includes bibliographical references and index.
ISBN 978-0-262-51580-1 (pbk. : alk. paper) 1. Architectural design. 2. Architectural design—Technological innovations. 3. Repetition (Aesthetics) 4. Design and technology. I. Title.
NA2750.C375 2011
720.1—dc22
2010031062

10 9 8 7 6 5 4 3 2 1

CONTENTS

PREFACE

Not long ago, in the nineties, no one doubted that a "digital revolution" was in the making—in architecture as in all aspects of life, science, and art. Today (early 2010) the very expression "digital revolution" has fallen into disuse, if not into disrepute; it sounds passé and archaic, at best the reminder of an age gone by. Yet digital technologies, now ubiquitous, have already significantly changed the way architecture is designed and made. They are changing how architecture is taught in schools, practiced, managed, even regulated. Etymologically, as well as politically, the notion of a revolution implies that something is or has been turned upside down. It may be too soon to tell if the digital is a revolution in architecture, but it is not too soon to ask what may be upended if it is. If the digital is a "paradigm shift," which paradigm is shifting? If architecture has seen a "digital turn," what course has turned?

This work will trace the rise of some aspects of modernity that have marked the history of Western architecture. They all relate to one key practice of modernity: the making of identical copies—of nature, art, objects, and media objects of all sorts. From the beginning of the Early Modern Age, and until very recently, the cultural demand and the technical supply of identical copies rose in sync. Identical copies inspired a new visual culture, and prompted new social and legal practices aimed at the protection of the original and its owner or creator. At the same time, new cultural technologies and new machines emerged and were developed to produce and mass-produce identical replications: from printed images and text set with moveable type to the industrial assembly line, from perspectival images to photography to the Xerox machine.

Two instances of identicality were crucial to the shaping of architectural modernity. The first was Leon Battista Alberti's invention of architectural design. In Alberti's theory, a building is the identical copy of the architect's design; with Alberti's separation in principle between design and making came the modern definition of the architect as an author, in the humanistic sense of the term. After Alberti's cultural revolution, the second wave of identical copies in architecture came with the industrial revolution, and the mass production of identical copies from mechanical master models, matrixes, imprints, or molds. Industrial standardization generates economies of scale—so long as all items in a series are the same.

The modern power of the identical came to an end with the rise of digital technologies. All that is digital is variable, and digital variability goes counter to all the postulates of identicality that have informed the history of Western cultural technologies for the last five centuries. In architecture this means the end of notational limitations, of industrial standardization, and, more generally, of the Albertian and authorial way of building by design.

This book recounts the rise and fall of the paradigm of identicality, and shows that digital and premechanical variability have many points in common. It discusses the rise of new forms of postindustrial digital craftsmanship by showing their relation to hand-making and to the cultures and technologies of variations that existed before the humanistic and modern rise of machine-made, identical copies. The first part of the book is a synopsis of the general argument; the second focuses on the mechanical rise and the digital fall of identical copies. A bit of repetition is inevitable, but the argument is simple—symmetrical, in a sense—with a beginning, climax, and end.

This chronicle situates today's computational tools in architecture within the ambit of a centuries-old tradition, with all of its twists and turns, of which the digital represents the most recent. Technologies change rapidly—"new" technologies in particular. To predict, and even interpret, new developments in cultural technologies on the basis of their recent history is risky, as one needs to extrapolate from a curve that is too short and build on evidence that has not been sifted by time. A more distant vantage point entails a loss of detail, but may reveal the outlines of more general trends. I shall endeavor to highlight some of these trends, and accordingly offer some conclusions—almost a morality, as in old tales.

In addition to the many friends, colleagues, and publishers that are mentioned in the book, special thanks are due to Megan Spriggs, who edited earlier drafts of most chapters, to Cynthia Davidson, who turned those chapters into a book, and to Peter Eisenman, who found a title for it.

VARIABLE, IDENTICAL, DIFFERENTIAL

On the evening of Sunday, August 15, 1971, U.S. President Richard Nixon announced in a televised speech[1] a series of drastic economic measures, including the suspension of a fixed conversion rate between the dollar and gold. The end of the gold standard, which had been reinstated by the Bretton Woods Agreement in 1944, had momentous economic consequences.[2] Its cultural fallout was equally epochal. Only a few years later, the founding fathers of postmodernism saw in "the agony of strong referentials"[3] one of the symptoms of the postmodern condition, and Nixon's abolition of the dollar's gold parity should certainly rank among the most prominent harbingers of many postmodern "fragmentations of master narratives"[4] to follow. From what is known of him, chances are that Nixon (who died in 1994) was never fully aware of his inspirational hold on Deleuze and Guattari's rhizomatic theories of mutability.[5] But from the point of view of historians of images, the end of the dollar-gold standard should also be noted for tolling the knell of one of the most amazing and miraculous powers that images ever held in the history of the West—one that art historians have often neglected.

British banking history may illustrate the relationship of paper currency and precious metal over a longer period of time than the history of the dollar would allow. From 1704, when banknotes were declared negotiable in England and Wales, until—with minor interruptions—1931, when the Bank of England in fact defaulted, any banknote issued by the Bank of England could be converted into gold or sterling silver at a fixed rate: paper stood for metal and one could be exchanged for the other at the same rate at any time. After Bretton Woods the British pound was pegged to the American dollar, and the dollar to gold, which, if one reads this story in British history books (and in Ian Fleming's *Goldfinger*),[6] means that the pound was once again on a gold standard, and if one reads it in American books means that the British pound was pegged to the dollar. Either way, the statement that still appears in small print on British banknotes—"I promise to pay the bearer on demand the sum of" £10, for example—before 1971 meant that the bearer would be paid on demand an amount of metal conventionally equivalent to ten pounds of sterling silver; as of 1971 and to this day, the same phrase means, somewhat tautologically, that the Bank of England may replace that banknote, on demand, with another one.[7]

The almost magic power of transmutation whereby paper could be turned into gold was canceled, apparently forever, on that eventful late summer night in 1971. For centuries before Nixon's intervention that alchemical quality of legal tender was guaranteed by the solvency of an issuing institution, but bestowed upon paper by the act of printing. For that miraculous power of images did not pertain to just any icon, but only to very particular ones: those that are identically reproduced, and are visually recognizable as such. Identicality and its instant visual recognition are what used to turn paper into gold; and identicality still makes legal tender work the way it does. A banknote that is not visually identical to all others in the same mass-produced series (with the exception of its unique serial number) may be fake or worthless. And as we have seen plenty of identical banknotes, until very recently we were expected to be able to tell at first sight when one is different, or looks strange, and reject it. Before the age of banknotes, the same pattern of visual identification applied to coins and seals, whose value and identification depended on the sheer indexicality of a mechanical imprint, and on the cultural and technical assumption that all valid copies could and would be reproduced identically.

These instances of "indexical" sameness—a quintessential feature of the mechanical age, and of mechanical reproducibility itself—are in direct contrast with other paradigms of vision that both preceded and followed the age of mechanical copies. To keep to monetary examples, the variability of artisanal handmaking survives today in personal checks, where the authority of the bank is attested to by the part of the check that is printed, but the validity of the check is triggered only by the manuscript signature of the payer. Like all things handmade, a signature is a visually variable sign, hence all signatures made by the same person are more or less different; yet they must also be more or less similar, otherwise they could not be identified. The pattern of recognition here is based not on sameness, but on similarity. Similarity and resemblance, however, are complex cognitive notions, as proven by the history of mimesis in the classical tradition, both in the visual arts and in the arts of discourse. Even today's most advanced optical readers cannot yet identify nor authenticate personal signatures, and not surprisingly, personal checks are neither universal nor standard means of payment (unless the bearer can be identified by other means, or is known in person, and trusted).

In the world of hand-making that preceded the machine-made environment, imitation and visual similarity were the norm, replication and visual identicality were the exception. And in the digital world that is now rapidly overtaking the mechanical world, visual identicality is quickly becoming irrelevant. Credit cards may well be in the shape of a golden rectangle (or a fair approximation thereof: it is not known whether this happened by chance or by design), and still bear logos, trademarks, and some archaic machine-readable characters in relief—a reminder of the time when they were invented in the late fifties. But today the validity of a credit card depends almost exclusively on a unique string of sixteen digits that identifies it, regardless of its format, color, or the material of which the card is made.[8] Indeed, for online transactions the physical existence of the card is neither required nor verifiable. The first way to confirm the validity of a credit card is to run a check on the sixteen-digit sequence of its number using a simple algorithm, known as Luhn's formula, which in most cases (statistically, nine times out of ten) is enough to detect irregularities. No one would try to judge the creditworthiness of a credit card by looking at it, in the way one would peruse a banknote or inspect its watermark. Visual identification is now out of the game. In this instance, exactly transmissible but invisible algorithms have already replaced all visual and physical traces of authenticity.

Albeit anecdotal, these monetary examples illustrate three paradigms of visual identification, essentially related to three different ways of making things. The signature, the banknote, and the credit card: when objects are handmade, as a signature is, variability in the processes of production generates differences and similarities between copies, and identification is based on visual resemblance; when objects are machine-made, as a banknote is, mass-produced, exactly repeatable mechanical imprints generate standardized products, and identification is based on visual identicality; when objects are digitally made, as are the latest machine-readable or chip-based credit cards, identification is based on the recognition of hidden patterns, on computational algorithms, or on other nonvisual features. This loss of visuality, which is inherent in the mode of use of the latest generation of credit cards, may in turn be a prelude to the eventual disappearance of the physical object itself: credit cards are in most cases already obsolete, as many of their functions may soon be taken over by cell phones, for example.

The list of objects of daily use that have been phased out by digital technologies is already a long one: digital consumer appliances tend to merge on a single, often generic technical platform a variety of functions that, until recently, used to be performed by a panoply of different manual, mechanical, or even electronic devices (from address books to alarm clocks to video players). Industrial designers and critics have taken due notice, as is shown by the ongoing debate on the disappearance of the object (or at least of some objects).[9] However, alongside and unrelated to this seemingly inevitable wave of product obsolescence—or perhaps, more appropriately, product evanescence—digital tools are also key in the design and production of a growing range of technical objects, old and new alike—from marble sculptures to silicon chips. And the technical logic of digital design and production differs from the traditional modes of manufacturing and machinofacturing in some key aspects.

A mechanical machine (for example, a press) makes objects. A digital machine (for example, a computer) makes, in the first instance, a sequence of numbers—a digital file. This file must at some later point in time be converted into an object (or a media object) by other machines, applications, or interfaces, which may also in turn be digitally controlled. But their control may be in someone else's hands; and the process of instantiation (the conversion of the digital script into a physical object) may then be severed in space and time from the making and the makers of the original file. As a consequence, the author of the original script may not be the only author of the end product, and may not determine all the final features of it.

To go back to image theory, a comparison may help to make the point. Each print of a picture in the same print run looks the same. All mass-produced series include minor accidental variances, but by and large, all buyers of the same postcard (printed, for example, in one thousand copies) will buy the same picture. On the contrary, a digital postcard, e-mailed from a computer to an electronic mailing list of one thousand recipients, is sent as a sequence of numbers that will become a picture again only upon delivery—when it appears on one thousand different computer screens, or is printed out by as many different printers. The digital file is the same for all. But each eventuation of that file (in this instance, its conversion into a picture) is likely to differ from the others, either by chance (some recipients may have different machines and applications), or by design (some recipients may have customized their machines or may deliberately alter the picture for viewing or printing). Some of this customizable variability certainly existed in the good old days of radio and television, and even of mechanically recorded music. But the degree of variability (and indeed, interactivity) that is inherent in the transmission and manipulation of digital signals is incomparably higher. We may well send the same digital postcard to all our friends. Yet there is no way to anticipate what each of them will actually see on the screen of his or her computer or cell phone (and even less, what they will see if they decide to print that picture on paper—or on any other material of their choice, for that matter).

The loss of visual significance that is so striking in the instance of the credit card may simply be the terminal phase of the general regime of visual variability—or sensorial variability if we include other senses beyond sight—that characterizes all digital environments. Variability is also a diacritical mark of all things handmade, but artisanal and digital variability differ in another essential feature. Handmade objects can be made on demand,

and made to measure. This makes them more expensive than comparable mass-produced, standardized items, but in compensation for their extra cost, custom-made objects are as a rule a better fit for their individual user. In other instances, however, artisanal variability may be a problem, rather than a solution. As hand-making is notoriously ill suited to delivering identical copies, this tends to be the case whenever identical copies are needed. To take an obvious example: before the invention of print the transmission of texts and images was at the mercy of the will and whims of individual copyists, who could make mistakes and unpredictable changes at all stages of the copying process. The inevitable random drift of all manually reproduced texts and images was for centuries a major impediment to the recording and the transmission of all sorts of cultural artifacts—from poetry and music to science and technology.

Some degree of randomness is equally intrinsic to all digital processes. In most cases, we don't know which machines will read the digital file we are making, or when, or what technical constraints or personal idiosyncrasies will ultimately determine the conversion of our work from machine-readable documents into something readable (or otherwise perceptible) for humans. But, to a much greater extent than was conceivable at the time of manual technologies, when every case was dealt with on its own merit, and individual variations were discussed, negotiated, and custom-made on demand, the very same process of differentiation can now be scripted, programmed, and to some extent designed. Variability can now become a part of an automated design and production chain. Indeed, this is what the most alert users of digital technologies have been doing for the last fifteen years or so—artists and technologists as well as entrepreneurs and capitalists.

Both the notions of a manual drift in artisanal and script cultures, and of a digital drift in contemporary computer-based environments, will be discussed at length in the central chapters of this book. But a simpler instance of digital "differentiality" (a term introduced by Greg Lynn to describe the new forms of serial variations in the digital age)[10] may clarify the matter here. As is well known, various features of many web pages are now automatically customized based on what the page makers know of each individual page user. This is why the advertising (and increasingly, the content) which appears on some of the most popular web sites differs based on the computer, the browser, network, or protocol we use to access those pages, and varies according to the time of day, the geographical location of the user, and a number of other arcane factors that are well-protected trade secrets. This is, at its basis, the golden formula that has made Google a very rich company.[11] Variability, which could be an obstacle in a traditional mechanical environment, where identical copies were pursued, expected, and had intrinsic value, has been turned into an asset in the new digital environment—indeed, into one of its most profitable assets. As content customization seems to be, for the time being, almost the only way to make digital content pay for itself, web users are learning to cope with its side effects. Readers of the same online edition of the same newspaper often end up reading, at the same time and in the same place, a permanently self-transforming hodgepodge of different texts and images (sounds can be added at will). Following on the same logic, experiments are reportedly underway to replace conventional printed billboards in public places with electronic ones, capable of detecting certain features of the onlookers standing in front of them (through physical or electronic markers) and adapting their content accordingly.[12]

There was a time when daily newspapers published more than one local edition (and a few still do); but the notion that each reader may find his or her own custom-made newspaper (or web portal, or advertisement in a railway station) to match his or her unique profile goes far beyond technical variability, or digital differentiality, and induces a feeling of cultural instability that many may find disturbing. Over the course of the last five centuries the "typographical man" became increasingly dependent upon a high degree of visual predictability to facilitate the storage and retrieval of written information. Visual and graphic stability in the layout of texts and images arose with print technology, and thence spread to all tools and instruments that were mechanically mass-produced (again, printed from the same matrix or mold). These same patterns of graphic recognition are still at the basis of many cultural and social practices that play an important role in the ordinary conduct of our daily lives. We used to look for a certain column (or index, or price) in the same place on the same page of the same newspaper; similarly, certain electromechanical interfaces, such as analog instrument panels with dials and gauges, used to assign specific data sources to fixed, distinct, and memorable visual loci (as in all cars of the same make, for example, where a given warning signal always lights up in the same place, form, and color on the dashboard).

None of this applies to digital interfaces, where even the fonts and sizes of alphabetical texts may change anytime, often without warning, and the same piece of information may pop up anywhere on the isotropic surface of a muted LED display or of an interactive control panel, in all kinds of different sensorial species (as sounds, pictograms, drawings, diagrams, alphabetical warnings in a variety of different languages, perfectly impenetrable numerical error codes, etc.). Indeed, there is a certain logic in that the company that most contributed to the variability of digital images (Adobe Systems, the makers of Photoshop) should also have created new software specifically to counter this digital drift—to freeze images and force users to view visually identical graphic layouts. Adobe's PDF, or "portable document format," essentially uses web technologies to transmit electronic photocopies—faxes sent over the Internet. Not without success: clearly, in many instances our societies cannot yet do without the iron inflexibility of the typographical page—a mechanical attribute *par excellence*. Tax forms must be identical for all (even when downloaded from a web site, or, more recently, filled in online) because line 33A-14 must appear on page 7 on all tax returns. This clearly shows how income tax returns could not have existed before the age of printing: even in the electronic era the internal revenue services of most countries, when they go online, are forced to use the most sophisticated technologies to reduce the ectoplasmic variations of digital images to the mechanical fixity of printed pages. The web sites of various ministries and national services that deal with tax returns are true works of electronic art, and Marshall McLuhan would have delighted in the digital emulation of Gutenberg's machine recently perfected by modern state bureaucracies: the typographical man is so integral to the modern state that the modern state, even after adopting electronic technologies, is forced to perpetuate a mimesis of the typographical world.[13]

So it seems, to sum up, that in the long duration of historical time the age of mass-produced, standardized, mechanical, and identical copies should be seen as an interlude, and a relatively brief one—sandwiched between the age of hand-making, which preceded it, and the digital age that is now replacing it. Hand-making begets variations, as does digital making; but the capacity to design and mass-produce serial variations (or differentiality) is specific to the present digital environment. Unlimited visual variability, however, may entail a loss of visual relevance: signs that change too often or too randomly may mean less, individually taken, and may in the end lose all meaning.[14] This was already the case in the old age of handmade variability, when the economy of visual communication was dysfunctional because of a penury of recognizable images, and is again the case in the new age of digital differentiality, where the economy of visual communication is dysfunctional because of an oversupply of variable images.

The sequential chronology of these three technical ages (the ages of hand-making, of mechanical making, and of digital making) lends itself to various interpretations.[15] Some objects were still handmade well into the mechanical age, and some will still be handmade, or mechanically made, well into the age of digital making. But, by and large, the second break in this sequence, the passage from mechanically made identical copies to digitally generated differential variations, is happening now. The first break, the transition from artisanal variability to mechanical identicality, occurred at different times in the past—depending on the classes of objects and technologies one takes into account. The defining shift from artisanal hand-making to mechanical manufacturing (or machinofacturing) came with the industrial revolution. However, if next to traditional objects of manufacturing (rails, sewing machines, or automobiles) we also look at media objects (texts, images, sounds, and their modes of recording and transmission), we may encounter some slightly different chronologies.

New media theorists[16] tend to situate the transition from variable to identical copies in the nineteenth or twentieth century, as they associate the rise of identicality with indexical realism, which is often seen as the distinctive property of photography and of cinema. Unlike an artist's drawing, a photographic image is a machine-made, quasi-automatic imprint of light onto a photosensitive film: by the way it is made, it can only record something that really happened. Traditional media scholars[17] relate the rise of identically reproduced, mechanical images to the invention of print and—almost simultaneously—of geometrical perspective in the Renaissance. Well before modern photographic technologies, Alberti first and famously defined perspectival images as the trace of light rays on a surface.

The history of architecture features a conflation of different technological timelines. Built architecture depends on the production of material objects (bricks, nails, iron beams, etc.), hence its modern history is linked to the traditional chronology of the industrial revolution. On the other hand, architectural design is a purely informational operation, and its processes are defined by a specific range of cultural and media technologies. For centuries the classical tradition was based on the recording, transmission, and imitation of architectural models. In turn, this tradition, or transmission, was and still is dependent on the media technologies that are available, at any given point in time, to record a trace of such models and to transmit them across space and time. What cannot be recorded will not be transmitted, and what is neither recorded nor transmitted cannot be imitated. Additionally, and unrelated to the publication, circulation, and reception of architectural rules and models, building may also be dependent on the cultural technologies needed to notate specific design instructions that are conceived by some to be carried out by others, sometimes in the absence of the original designer. A key issue in the modern, notational theory of architectural design, this technical, point-to-point exchange of building and construction data is once again a matter of recording and transmission—a media problem.

1.1 Architecture and the Identical Copy: Timelines

The history of architecture in the machine age is well known. As it has been written and rewritten many times over by the militant historians of twentieth-century modernism and by their followers, it is a tale of sin and redemption. Architecture was slow in coming to terms with the industrial revolution. Throughout the nineteenth century, most architects either ignored or reacted against the new technologies of industrial mass production. Then came the pioneers of modern architecture, and their wake-up call. As Le Corbusier and others began to claim in the early twenties, mechanization was changing the world, and architecture had to rise to the challenge. Architects should invent new architectural forms, made to measure for the new tools of mechanical mass production; and town planners should invent new urban forms, made to measure for the new tools of mechanical mass transportation. For the rest of the twentieth century many architects and urbanists did just that. Oddly, many architects and urbanists are still doing that right now, as they ignore, or deny, that today's machines are no longer those that Le Corbusier and his friends celebrated and sublimated almost a century ago.

Well before the industrial revolution, however, another mechanical revolution had already changed the history of architecture. Printed books are a quintessentially industrial product. They are mass-produced. Mass production generates economies of scale, which makes them cheaper than manuscript copies. They are standardized—each copy is the imprint on paper of the same mechanical matrix. Early modern printed books were so much cheaper and better than coeval handmade books that they soon replaced them in all markets, and the new architectural books in print (manuals, treatises, pattern books, etc.) changed the course of architecture first and foremost because of the printed images they contained. Before the invention of print, manual copies of drawings were famously untrustworthy, and as a result, images were seldom used, or altogether avoided, whenever precise copies were required. In such cases, nonvisual and prompted a culture of identical copies that became pervasive in the West well before the industrial revolution, and the actual rise of mechanical mass production. Standardized images preceded industrial assembly lines, and a culture of standardized architecture was already well established at a time when all visually standardized architectural parts (from moldings, columns, and capitals to windows, chimneys, etc.) had to be carefully handcrafted in order to *look* identical to one another.[20] In the process, standardized images standardized the craftsman's gesture: the free hands of artisans were coerced to iterate identical actions, working like machines that at the time no one could imagine or presage—but which would eventually come, churning out identical copies better and cheaper than any artisan could or would. This is where modern Taylorism and mechanization took over, Le Corbusier stepped in, and the second, and better-known part of the story began.

1.2 Allography and Notations

As it happens, at the very moment printed images were revolutionizing the transmission of architectural models, another media revolution was crucially changing the way architects work. Alongside the images of eminent buildings of the past or present, and the new sets of ready-made visual models that would characterize early modern architectural books in print, another class of architectural drawings and models was fast rising to prominence: the project documents that Renaissance architects produced in growing numbers and forwarded to increasingly distant building sites—a physical distance that went hand in hand with the growing intellectual and social estrangement between architects and builders. New reproduction technologies were of no consequence for project drawings, as these technical documents destined for builders were not meant to be mass-produced: each drawing could be hand-drafted as precisely as needed before being shipped to the site where it would be used, without any loss in precision or other risks that would have come with copies. The only technological innovation in Renaissance project drawings may well have been their very invention—or the invention of their mode of use.

According to Nelson Goodman, all arts were born autographic—handmade by their authors. Then, some arts became allographic: scripted by their authors in order to be materially executed by others.[21] When did architecture evolve from its pristine autographic status as a craft (conceived and made by artisan builders) to its modern allographic definition as an art (designed by one to be constructed by others)? The traditional view, which attributes to early modern humanism the invention of the modern architect, and of his new professional role, rests upon some famous narratives: Brunelleschi's legendary struggle for the recognition of his role as the sole conceiver and master of a major building program; Alberti's radical claim that architects should be not makers but designers, and his definition of a modern notational system of scaled architectural drawings in plan and elevation that were the indispensable means to this end.

Counter to these clear-cut stereotypes, it is easy to point out that the separation between design and building (and between designers and workers) is a matter of degrees. Architectural notations of some kind have almost always existed. It seems that at the beginning of historical time Egyptian architects already used fairly precise architectural construction drawings.[22] But the history of design processes in antiquity is a difficult and controversial subject, as archaeological scholarship on the matter must build on slender evidence. Indeed, the evidence is at times so thin that some archaeologists have concluded that Greek architects of the classical age did not use scaled drawings at all; other known ancient notational systems, such as textual instructions, three-dimensional models, templates, or full-size diagrams, sometimes incised on stones or walls or otherwise sketched on site, all imply or require some presence of the designer on the site of construction. The use of scaled project drawings would have arisen only in the Hellenistic period, alongside the growing estrangement between designers and craftsmen that the introduction of a more advanced notational system suggests.[23] The controversy is compounded by the ambiguity on this issue of the most important extant source, Vitruvius's treatise. Vitruvius's famously obscure definitions of three kinds of architectural drawings (*ichnographia, orthographia, scaenographia*) in his first book seem to take some practice of architectural drawings for granted. Yet his own design method never refers to, and does not require, any kind of scaled drawing.[24]

Beside archaeologists and classical scholars, medievalists have also weighed in on the matter. Something similar to proportionally drawn plans and elevations can be dated to the thirteenth century, and more convincingly to the fourteenth and fifteenth centuries (the famous drawings from the workshop of Peter Parler in Prague are coeval to Brunelleschi's work on the Florence dome). These, and other textual documents, have led some to suggest that "construction by remote control" was common among Gothic master builders, and that adequate notational tools, and social practices, already existed to support such design methods well before, and unrelated to, the new architectural theory of the Italian humanists. This thesis has been corroborated by an unusual blunder by the eminent scholar Wolfgang Lotz, who in his seminal 1956 essay on Renaissance architectural drawings misread a crucial passage in Alberti's *De re aedificatoria*, wrongly concluding that Alberti encouraged architects to draw in perspective, and that Raphael's "Letter to Leo X" (1519), rather than Alberti's treatise, should be credited with the modern "definition of the orthogonal projection."[25] Although Lotz eventually corrected himself, that essay is one of the sources of a persistent tradition according to which the pictorially oriented Renaissance architects of the South, far from having developed the "orthogonal" notational format, would in fact have delayed its rise due to their penchant for perspectival, illusionistic, nontechnical drawings.[26]

Recent scholarship has pointed out that Raphael's passages on architectural drawings are little more than an amplification of Alberti's theory on the matter,[27] but the idea that "orthogonal projections" may have been invented by Gothic builders, or even by Renaissance architects, is problematic on other counts. Orthogonal, or parallel projections, as defined by Gaspard Monge's descriptive geometry (1799), posit a center of projection located at infinity (the only possible point of origin for rays, beams, or vectors, that must all be parallel to each other on arrival): in today's projective geometry, central and parallel projections differ only in that the projection center is a proper point for the former, and an improper point (i.e., a point at infinity) for the latter. The drawing of "orthogonal" ground plans may not require projections of any kind, as the ground plan of a building may simply be seen as its imprint or trace on a real site (if necessary, redrawn to scale). But "orthogonal" front views, or elevations, are a trickier matter.

According to late medieval optics, and to Alberti's own geometrical perspective, "orthogonal" front views would have required an observer's eye to be physically pushed back to an infinite distance, which, as a Renaissance mathematician famously remarked, is actually "nowhere."[28] Late medieval and early modern geometries, owing to their Aristotelian framework, did not allow for such insouciant appropriations of infinity, nor could they supply the homogeneity and continuity of space that parallel projections from infinity would have demanded. Piero della Francesca drew at least one famous head in plan—actually, two plans, an elevation, and a side view, where all the views are connected by parallel projection lines, more than three centuries before Monge's *Descriptive Geometry*.[29] Likewise, late medieval and Renaissance architects used, and depended upon, simpler sets of "orthogonal" plans, elevations, and side views (and later, sections) which, however, no mathematician at the time could have defined, nor formalized, for lack of any workable notion of geometrical infinity. Parallel projections were for centuries a practice without a theory.[30]

Yet in this matter too Alberti scored a major breakthrough. Alberti could, and did, codify central projections, which represent infinity (as a vanishing point) without defining it; but he could not codify parallel projections, which would have posited a physical eye (the perspectival point of view) in a nonphysical place (infinity). However, precisely because he had already defined central projections in his treatise on painting, when a few years later he wrote his treatise on architecture Alberti could for the first time ever lay out precisely what architects should *not* do: architects should avoid perspective, as from foreshortened lines one cannot take precise measurements (in Raphael's slightly later wording). As Alberti mandates in a key passage in the second book of *De re aedificatoria*, architects' drawings, unlike painters' perspectival views, require "consistent lines," "true angles," and "real measurements, drawn to scale."[31]

One needs perspective to have been invented in order to tell architects not to use it. As a side effect of his invention of geometrical perspective, Alberti could provide the first (albeit negative) geometrical definition of modern proportional and orthogonal plans and elevations—at a time when geometry did not allow for any definition of parallel projections. This may appear to be a fine point of geometry (and it is, as it is tantamount to defining parallel projections as "noncentral" projections, without a corresponding center of projection at infinity); but, at a more practical level, Alberti's strategy was also consistent with the basic need to explain why scaled elevation drawings should not include foreshortened lines (as such drawings often did before Alberti, and occasionally kept doing after him).

Alberti's distinction between building and design (*lineamenta*) is spelled out in various but unequivocal terms in the first, second, and ninth books of *De re aedificatoria*, and it is one of the foundational principles of his entire architectural theory.[32] His new geometrical definition of architectural project drawings (and models) provided a consistent set of notational tools suited to his new, allographic way of building. As previously mentioned, the distance between designers and building sites is an historical variable, and it ebbed and flowed for centuries before (and after) Alberti's theoretical climax. With these ebbs and flows, the need for, and availability of, reliable notational tools varied over time, and the evidence that several ways of building by notation existed, and were variously implemented, before Alberti should not be discounted. But, in addition to the sharpness of its conceptual proclamation, the Albertian way differed from all precedents in another, essential aspect—one that has stayed to this day.

1.3 Authorship

As Alberti repeatedly emphasizes in his treatise, architects must work with drawings and three-dimensional models throughout the design process, as various aspects of the project cannot be verified unless they are visualized.[33] Drawings (albeit, apparently, not models)[34] will also be used as notational tools when the project is finalized and construction drawings are sent to the builders; but the two functions, visualization and notation, remain distinct. Designers first need drawings and models to explore, nurture, and develop the idea of the building that, as Alberti states at the outset of his treatise, is "conceived in the mind, made up of lines and angles, and perfected in the learned intellect and imagination."[35] Alberti insists that models should also be used to consult experts and seek their advice; as revisions, corrections, and new versions accumulate, the design changes over time; the whole project must be examined and reexamined "not two, but three, four, seven, ten times, and taking breaks in between."[36] The final and definitive version is attained only when each part has been so thoroughly examined that "any further addition, subtraction or change could only be for the worse."[37] This is when all revisions stop, and the final blueprint (as we would have said until recently, both literally and figuratively) is handed over to the builders. Thenceforward, no more changes may occur. The designer is no longer allowed to change his mind, and builders are not expected to have opinions on design matters. They must build the building as is—as it was designed and notated.

At various times and in different contexts Alberti insists on this ideal point of no return, where all design revisions should stop, and construction begin speedily and without hesitation (and, he adds, without any variation or change during the course of the works, regardless of who is in charge of the site).[38] Alberti famously advised architects against directing the actual construction: in his view, building should be left to the workers and to their supervisors.[39] He allows that "to have others' hands execute what you have conceived in your mind is a toilsome business,"[40] and indeed documents related to the building history of the Tempio Malatestiano in Rimini[41] prove that the throes of allography did not fail to take a toll on Alberti's career as an architect. Local workers, craftsmen, and master builders might not have measuring them (and Alberti himself implies they should not be, presumably for the same reason). Alberti's apparent preference for drawings, rather than models, as the primary notational tool marked a significant departure from the late medieval tradition.[43] But drawings, unlike buildings and models, are two-dimensional, and in most cases proportionally smaller than the building itself.[44] Consequently, a building and its design can only be *notationally* identical: their identicality depends on a notational system that determines how to translate one into the other. When this condition of *notational identicality* is satisfied, the author of the drawing becomes the author of the building, and the architect can claim some form of ownership over a building which in most cases he does not in fact own, and which he certainly did not build—indeed, which he may never even have touched. The transition from Brunelleschi's artisanal authorship ("this building is mine because I made it") to Alberti's intellectual authorship ("this building is mine because I designed it") is discussed in more detail in section 2.6 below. The notion of an architect's intellectual "ownership" of his work is not spelled out by Alberti in so many words, but it is inherent in the notion of authorship that Alberti borrowed from the humanists' arts of discourse and applied, for the first time ever, to the art of building.

Thus reformed, architecture ideally acquires a fully authorial, allographic, notational status. Insofar as a building and its design are considered notationally identical, one can identify an architectural work either with the design of the building or with the building itself (a step that Nelson Goodman still hesitated to take in 1968).[45] Around 1450 Alberti's claim to architectural authorship (as well as his new way of building by notation) must have appeared outlandish or worse, culturally as well as technically. Yet Alberti's authorial ambitions and concerns were common among writers, rhetoricians, and scholars of his time.

Alberti, the humanist, was painfully aware of the inevitable destiny that awaited all texts and images when severed from the hands of their authors and caught in the unpredictable drift of scribal production. Catullus could well "smooth with dry pumice" his brand-new papyrus roll of poems before presenting it to his first, and possibly fictional, dedicatee; but that original finishing touch was but a frail seal, and fifteenth-century humanists knew that most extant classical texts were mosaics of citations, interpolations, additions, subtractions, and plain copy errors. Modern philology was developed precisely to reconstruct, as much as possible, the original text of the author—the one the author would have "smoothed with dry pumice" on the day he considered his manuscript finished. Contemporary philologists and linguists have also suggested that in the late Middle Ages the awareness of the technical variability (*mouvance*, or drift) of scribal copies prompted new modes of textual interaction, where variances were not only tolerated, but actually expected, encouraged, and sometimes exploited.[46] The very notion of an "original" would hardly apply in such a context, as the so-called originals would be too many, and none more relevant than any other.

There is additional evidence that some of the early humanists (Poggio Bracciolini in particular) availed themselves of this potentially interactive format to circulate manuscripts that invited feedback, comments, and additions.[47] Alberti himself may have engaged in this practice—and indeed, it would be fascinating to see Poggio and Alberti as active wikipedists of the late scribal age.[48] This process of multiple revisions of, and possibly interactive feedback on, the successive drafts of a literary text corresponds to the fluid state of architectural design during the "versioning" phases of its development, on which Alberti insists so emphatically.[49] But Alberti also evidently thought that when revisions stop, they should stop for good—and forever. Alberti was so anxious about scribal errors that he took the unusual step of flagging passages where he thought that copyists of his manuscripts might be more at risk of being led astray, and devised some ploys (as well as some full-fledged and bizarre devices) to limit that risk, and contain potential damage.

A wealth of evidence proves that, when the final version of a given text was attained, Alberti aimed at having it copied and reproduced as faithfully as possible, and with as little external intervention as was conceivable at the time. The best technical means to this end would have been for Alberti to have his texts and illustrations printed—which, however, he could not or would not do, mostly for chronological reasons, although he may have considered the option toward the end of his life. Indeed, Alberti's pursuit of identical copies is exactly coeval to the development of print technologies, and this parallel chronology is certainly not a coincidence. Alberti's insistence on an ideal, but drastic, authorial cutoff—the point at which all revisions stop and identical replication starts—curiously anticipates a practice that eventually became common in the printing industry, and survives to this day in the technical term *bon à tirer* (good to print).[50] Originally, the author's *bon à tirer* (normally dated and signed) written on the last proofs validated the final version of his or her text, and "authorized" its identical replication in print. Thenceforward, readers could expect exactly the same words in each copy as in the author's original, even though the author never printed, nor necessarily signed, any individual book. Thanks to the cultural and technical logic of mechanical replication, authorship was extended from the author's original to all identical copies of it.

For intellectual and ideological reasons, which should be seen in the context of the humanists' invention of modern authorship, and perhaps in the larger context of the humanists' contribution to the shaping of the modern self and of the notion of individual responsibility, Alberti anticipated this division between an author's work and its mechanical reproduction. But Alberti tried to impose this authorial paradigm within the ambit of a manual production chain, where no machine would deliver identical copies, and scribes could be reasonably expected to produce just the opposite—randomly changing, individual variations. Also, but crucially for the history of architecture, Alberti extended his precocious *bon à tirer* paradigm from literary to architectural authorship, asserting that the same conditions and the same consequences should apply. The fact that in most cases the architect's design should beget only one building (and not a series of copies, as would a printing press, or a late medieval scriptorium) is irrelevant in this context.[51] What matters is the relation of identicality between the original and its reproduction. Alberti's entire architectural theory is predicated on the notational sameness between design and building, implying that drawings can, and must, be identically translated into three-dimensional objects. In Alberti's theory, *the design of a building is the original, and the building is its copy*.

1.4 The Early Modern Pursuit of Identical Reproduction

Between the end of the Middle Ages and the beginning of the Modern Age, two almost simultaneous media revolutions changed the course of European architecture. On the one hand, print transformed the modes of transmission of architectural information in space and time. For the first time in the West, texts and images could be protected from the permanent drift of scribal transmission and frozen as prints—mechanically reproduced, identical copies. On the other hand, a new notational format was then starting to reshape the transmission of architectural data from designers to builders: a development related to the rise of new forms of allography in building, and to the growing gap between thinkers and makers.

These changes in architectural notation were not related to the development of any printing technology, as they did not require any.[52] Yet right from the start the spirit of the Albertian design process, which aimed at the identical materialization of the architect's design of a building, was already, in a sense, mechanical. In today's terms, Alberti's authorial way of building by notation can be interpreted as an ideally indexical operation, where the architect's design acts like a matrix that is stamped out in its final three-dimensional result—the building itself. This metaphor may seem far-fetched, but it will sound familiar to those acquainted with today's tools for three-dimensional digital fabrication.[53]

The indexical nature of the Albertian design process also resonates with the design theory of one of the most prominent architectural thinkers of the late twentieth century, Peter Eisenman. Eisenman's theory of indexicality in design stems from the same premises as Alberti's. In both cases the authorial mark is inscribed in the project, and its expression in a constructed object matters only insofar as the end product is the identical trace (or index) of its conceptual matrix—all variation being irrelevant or erroneous. And in both cases, albeit in Eisenman's somewhat more deliberately, the built work can be seen as a probe or critique of the limits of allographic authorship. Given his pivotal role in the digital turn of the nineties, Eisenman's work is the ideal touchstone to assess the continuing relevance of the Albertian paradigm to the theory and practice of contemporary design.[54] This is even more true if, as I shall argue, the Albertian paradigm is now being reversed by the digital turn. A paradigm must be asserted prior to being reversed.

Alberti's new way of building by notation should be seen in the context of the quest for identical replication that is at the core of Alberti's work and theory, in all the diverse fields he tackled. And identicality is the common denominator to all new cultural technologies that crossed paths with, and transformed, the art of building in the Renaissance. In the case of books, identical copies in print were obtained from a mechanical matrix; in the case of building, identical construction was obtained from a notational matrix. The former process was induced by a new mechanical technology: print. The latter, albeit similar in its mechanical spirit, depended entirely upon cultural conventions: a reformed social practice (authorship), and a new cultural technology (a reliable notational format, or protocol, for architectural project drawings).

1.5 Geometry, Algorism, and the Notational Bottleneck

Alberti's design process relies on a system of notation whereby all aspects of a building must be scripted by one author and unambiguously understood by all builders.[55] Its principal notational means reside in the scaled and measured drawings of plans, elevations, and side views defined in the second book of *De re aedificatoria*. As mentioned above, Alberti's definition of such drawings as nonperspectival led him to describe them in terms that anticipate, for most practical purposes, today's theory of orthogonal or parallel projections.

The development of this new notational format was accompanied by a drastic transformation in the nature and function of the architect's mathematical tools. Vitruvius's design method was based on proportional modular systems, where each modular unit was a constituent part of the building, aptly chosen to be easily discernible and measurable while the building itself was being built.[56] The dimensions of most other parts were then defined as multiples or fractions of one of these modules, but except for the most straightforward cases (for example, "eight modules," or "half a module") these ratios were not indicated as numbers. Instead, Vitruvius narrated sequences of geometrical constructions that could be carried out "live," so to speak—on site and at full scale, with rulers and compasses. These mechanical operations could determine the real size of all relevant parts of the building without any need to measure them (except for the first one, the module itself), and crucially, without any need to follow scaled or measured construction drawings.

Vitruvius's design method may already have been outdated at the time of his writing (and was possibly inconsistent with other parts of his own theory).[57] Regardless, this is the method of quantification that Vitruvius, with his almost undisputed authority, bequeathed to Renaissance and early modern theoreticians. Not surprisingly, Renaissance architects soon found Vitruvius's text-based, narrational, and formulaic geometry odd and, at times, unwieldy. Vitruvius's modular system may have been a speedy way to *make* things (on site), but it was a cumbersome way to *design* them (off site). With the rising popularity of the modern way of building by design, and the concomitant rise of Hindu-Arabic numeracy in the West (then called "algorism"), Renaissance theoreticians gradually adapted Vitruvius's autographic and, in a sense, artisanal construction process to the new format of scaled and measured project drawings. The conversion from Vitruvius's geometrical constructions to modern computation was slow and laborious, but in time number-based operations phased out the manual, artisanal practice of ruler and compasses, and architectural measurements ceased to be geometrically determined by impromptu diagramming and real-size, on-site tracing and stonecutting.[58] Instead, it became increasingly common to expect that all dimensions should be precalculated and the drawing, because in that case their scales change. Architects soon realized that the new way of building from scaled drawings had a catch: if you cannot measure an object in a drawing, then no one can build it.

If all that is built is built from notations, and if the drawings (or models) must contain all of the necessary data for an object to be built identically to its design, it follows that in most cases what can be built is determined by what can be drawn and measured in drawings. And as the notational system that encodes and carries data in architectural design is mainly geometric, it also follows that the potency of some geometrical tools determines the universe of forms that may or may not be built at any given point in time (with some nuances based on costs and on the complexity of the geometrical operations).

This notational bottleneck was the inevitable companion of all allographic architecture from its very start. Forms that are difficult to draw and measure used to be difficult or impossible to build by notation. Robin Evans has shown how some well-known architects tried to dodge the issue. Parts of Le Corbusier's church at Ronchamp, for example, were meant to look like plastic, sculptural, and irregular volumes—hand-shaped, like the sketches and three-dimensional models from which they were derived. Behind the scenes, though, Le Corbusier's engineers had to cook the books so that the most sculptural parts of the building could be duly drawn and measured in orthogonal projections. The roof in particular was redesigned as a regular, albeit sophisticated, ruled surface. This high-tech geometrical construction was accurately and laboriously devised to approximate Le Corbusier's supposedly instinctive, unscripted gesture as closely as possible. Evans also suggests that Le Corbusier was aware of and complicit in this ploy.[61]

In the most extreme cases, when a form is too difficult to notate geometrically, the last resort of the designer may well be to abandon the modern design process altogether, and return to the traditional, pre-Albertian, autographic way of building. If you can't draw what you have in mind in order to have others make it for you, you can still try to make it yourself. For example, this is what Antoni Gaudí did, most famously in the church of the Sagrada Familia, not coincidentally reviving, together with architectural forms evocative of a Gothic cathedral, some of the technologies and the social organization of a late medieval building site.[62] Gaudí built some parts of the Sagrada Familia much as Brunelleschi had built his dome in Florence: without construction drawings, but supervising all and everything in person, as an artisan/author who explains *viva voce* or shapes with his hands what he has in mind. It is not by chance that Gaudí is a famous case study among contemporary digital designers: once again, new digital tools and preallographic, artisanal fabrication processes find themselves, sometimes unintentionally, on similar grounds.[63]

This apparent affinity between manual and digital technologies is further evidence of a deeper and vaster connection, which will be further discussed in the second part of this book. But from this historical narrative also follows another, preliminary but inescapable remark. Since the establishment of the modern, allographic way of building, a notational mediator has stood between the ideas of the architect and their expression in building. For centuries, this mediator was essentially geometric: architects had to use two-dimensional drawings to script the forms of three-dimensional objects. They did so using the conventions and under the constraints of a geometrical language that, like all languages, was never universal or neutral. Then came computer-aided design.

Early in the history of computer-aided design (the actual chronology varies with the development and releases of specific families of software) architects started to realize that, even though a computer screen is two-dimensional, all three-dimensional forms visualized through it may exist in a computational three-dimensional space right from the start. Regardless of the interfaces and the conventions chosen to represent them, all geometrical points controlled by recent 3D CAD or animation software are, at their root, a set of three coordinates that locate each point in a three-dimensional space. As a result, a coherent object designed on a computer screen is automatically measured and built informationally—and the computer can actually fabricate the same object for good, if necessary, via a suitable 3D printer.

Indeed, 3D printing, 3D scanning and reverse modeling have already made it possible to envisage a continuous design and production process where one or more designers may intervene, seamlessly, on a variety of two-dimensional visualizations and three-dimensional representations (or printouts) of the same object, and where all interventions or revisions can be incorporated into the same master file of the project. This way of operating evokes somehow an ideal state of original, autographical, artisanal hand-making, except that in a digitized production chain the primary object of design is now an informational model. The range of its possible eventuations, in two and three dimensions, at all scales, and in all formats, includes the fabrication of the object itself.

By bridging the gap between design and production, this mode of digital making also reduces the limits that previously applied under the notational regimes of descriptive and predescriptive geometries, and this may well mean the end of the "notational bottleneck" that was the uninvited guest of architectural design throughout most of its early modern and modern history. Under the former dominion of geometry, what was not measurable in a drawing was not buildable. Now all that is digitally designed is, by definition and from the start, measured, hence geometrically defined and buildable.[64] Yet a cautionary note may be in order here.

For all of its almost unlimited versatility, the computer is still a tool—a technical mediator that in this instance is interposed between a designer and an object of design. All tools feed back onto the actions of their users, and digital tools are no exception. All design software tends to favor some solutions to the detriment of others, and as a consequence most digitally designed or manufactured objects can easily reveal their software bloodline to educated observers. However, the scope of these new constraints should be seen in light of the old ones, which held sway for centuries.

Since its inception, the notational regime of geometry imposed upon architects a strict diet of straight lines, right angles, squares and circles, and some bland variations on similarly elementary Euclidean themes. The few significant exceptions that have marked the history of architecture were realized, for the most part, nonallographically (that is, in part or entirely without the mediation of scaled construction drawings). In 1925, Le Corbusier published an actual synopsis of primary-school geometry (a table of lines, regular surfaces, and elementarysolids as found "on the back of exercise books issued to the elementary schools of France"), proudly stating: "this is geometry."[65] The repertoire of forms available to architects today is so vast as to appear unlimited, and it includes nongeometrical forms (sometimes also called "free forms"), which can now be digitally scanned, measured, and built. Evidently, the old notational bottleneck has not disappeared; but for most practical purposes digital technologies have already made it almost unnoticeable, and often irrelevant.

Indeed, in many cases, today's digital designers are no longer working on notations of objects, but on interactive avatars (or informational models) of the objects themselves. Digital technologies for design and fabrication may in such cases still be seen as instrumental mediators, but functionally they are more akin to material utensils, like hammers and chisels, than to traditional notational vectors such as blueprints or construction drawings. CAD-CAM applications are responsive tools for designing and making at the same time, not recording tools for scripting a final but inert set of design instructions.

1.6 The Fall of the Identicals

In his first book, published in 1970, Nicholas Negroponte prophesied an "Architecture Machine" that could act as an all-purpose cybernetic design assistant, and make possible through digitally mediated collaboration a high-tech version of Bernard Rudofsky's "architecture without architects."[66] Remarkably, in 2009, some of the latest trends in digital design seem to hark back to Negroponte's earliest anticipations. On the purely technical side, distributed or "cloud" computing recalls aspects of the mainframe environment that Negroponte would have been familiar with in the sixties and seventies. And recent developments in information modeling software are giving new prominence to the collaborative, information-based, decision-making aspects of the design process, which had been jettisoned by the more tectonically oriented CAD-CAM technologies of the nineties.[67] Whether through revival or survival, some of the vintage cybernetic "architecturology" of the seventies appears to be staging a comeback—an odd vindication of sorts for a generation of prophets who, until recently, appeared to have gotten it all wrong. For when the digital revolution arrived for good, in the eighties (and, for architecture, in the nineties), it took a turn that none of its early advocates had anticipated.

In January 1982, *Time* magazine proclaimed the IBM PC to be "Man of the Year." Ten years later, in the fall term of 1992, Columbia University's Graduate School of Architecture, Planning and Preservation inaugurated its seminal "paperless studio." But the cultural relevance of this factual chronology is debatable. It is a well-known pattern in the history of technosocial change that new and potentially disruptive technologies are often first tasked to emulate preexisting ones.[68] Indeed, many in the early nineties (including some distinguished technologists) were persuaded that CAD software would serve primarily to make cheaper, faster renderings and project drawings—easier to edit, archive, and retrieve.

The idea that the new digital design tools could also serve to make something else—something that would not otherwise have been possible—may have occurred when architects began to realize that computer-aided design could eliminate many geometrical and notational limitations that were deeply ingrained in the history of architectural design. Almost overnight, a whole new universe of forms opened up to digital designers. Objects that, prior to the introduction of digital technologies, would have been exceedingly difficult to represent geometrically, and could have been produced only by hand, could now be easily designed and machine-made using computers. Perhaps, some claimed, too easily.

As a side effect of this technological upheaval, complex or irregular geometries, which throughout most of the twentieth century often stood for some form of technological aversion on the part of the architect (because they could not be geometrically notated, nor machine-made, but had to be laboriously handcrafted), suddenly acquired the opposite meaning. Intricate, knotted, and warped forms became a trademark of the new digital tools, and signs of a new wave of excitement about technological change. The nineties (like the twenties) were a decade of technological optimism. Some critics failed to take notice, and as a result many exuberantly irregular, digitally made forms of the nineties have been described as "expressionist"—uncanny or anxious.[69]

One of the most acclaimed digital designers of the nineties may have unintentionally contributed to this critical misunderstanding. More than anyone else's, Frank Gehry's buildings of the time (particularly the Guggenheim Bilbao) brought the digital turn to the architectural forefront, as a stunned and often admiring general public concluded that digital technologies were indeed triggering an architectural revolution. This may have been true, but in Gehry's case, appearances were misleading. As is well known, Gehry's design process at the time began with handmade, sculptural models. These were then handed over to a technical team to be converted into geometrical drawings, as per the good old notational paradigm.[70] As mentioned above, a similar situation in Le Corbusier's office in the early fifties (1952–1954) ended in a standstill, and Le Corbusier's engineers had to alter the original model for Ronchamp in order to make it geometrically measurable. But by the early nineties (1991–1994; the Guggenheim Bilbao was inaugurated in 1997), thanks to digital technologies, the geometrical representation of irregular (or "free-form") three-dimensional objects had become a relatively easy task.

At the time, several tools (some derived from medical instruments) were already available to scan and digitize all kinds of objects, regardless of their form, or formlessness. First, physical models had to be converted into their digital doppelgängers by scanning a sufficient number of their surface points. The digital process of design and manufacturing could then take over. After much further work (on the digital model as well as on new drawings and physical models and prototypes derived from it) the process culminated in the ideal "printout" of the digital model—at the real scale of the actual building. In practice, the final construction was much more laborious, as it involved the three-dimensional assembly of a large number of digitally fabricated components. In theory, however, digital technologies in this instance acted as little more than a virtual three-dimensional pantograph. They were used to measure a three-dimensional prototype and replicate it identically at another, usually enlarged, scale. The reference to Christoph Scheiner's pantograph is not metaphorical. Alongside his better-known planar pantograph, Scheiner had also devised a spatial one, which, however, he stopped short of applying to the homothetic magnification of three-dimensional objects. No stereographic pantograph seems ever to have been used for architectural purposes—before Frank Gehry, that is.[71]

Digital tools in Gehry's office were used to further, not to transcend, the architects' traditional pursuit of identical replications. For centuries, project drawings had to be laboriously translated into notationally identical constructed objects. Gehry's engineers could do this faster and better than their predecessors; apparently, they could notate project measurements straight out of three-dimensional models (an unprecedented feat), and measure, then fabricate, some very ungeometrical surfaces. For all of its complexity, this was an allographic strategy that Alberti could have understood, if not praised. Gehry's pantographical process did not mark the end, but the climax of the notational paradigm—carried over, through digital tools, from an older world of simpler geometries into a new universe of "free forms" and unprecedented formal complexity.

Nothing prevents digital technologies from being used to make identical (or homothetic) copies. Indeed, anyone can use a computer with a scanner and printer to emulate a photocopier. But this is neither the smartest nor the most cost-effective way to use a computer. Concurrent with the construction of the Guggenheim Bilbao, new theories were emerging to claim just that—namely, that digital technologies could be put to better use designing and building digitally variable objects, rather than making three-dimensional copies; and that digital design could be digital from the start (i.e., design could start from algorithms rather than from the scanning and scaling of physical models). As it happens, the discourse on digital variability in architecture was sparked, in the late eighties and early nineties, by a most unlikely conflation of thinkers and ideas: the seventeenth-century philosopher and mathematician Gottfried Wilhelm Leibniz; Gilles Deleuze's book *The Fold: Leibniz and the Baroque* (1988; and 1993 in translation); Bernard Cache's contribution to, and subsequent interpretation of, the latter; and Peter Eisenman's and Greg Lynn's creative adaptation of the Deleuzian fold to American postdeconstructivist architectural theory. These disparate sources somehow came together, blended and fused in a special issue of *Architectural Design*, "Folding in Architecture," published in 1993.

There were many reasons why Leibniz's mathematics of continuity should appeal to digital designers of the early nineties: the design software of the time could easily manipulate continuous functions, thus putting Leibniz's differential calculus within the reach of most architects, regardless of their mathematical skills; and the earlier devices for numerically controlled fabrication could mill or mold or otherwise print out a vast range of continuous and curvaceous lines with great facility and at little cost. Additionally, Deleuze's often nebulous definitions of the "fold" (originally, a point of inflection in a continuous function) and Deleuze and Cache's descriptions of the "objectile" (originally, the notation of a parametric function) were more enthralling than the mathematical formulas from which they derived. Without Deleuze's timely mediation, few architects would have found high school calculus so highly inspiring. Regardless, Deleuze's and Cache's objectile ranks to this day among the most apt definitions of the new technical object in the digital age: the objectile is not an object but an algorithm—a parametric function which may determine an infinite variety of objects, all different (one for each set of parameters) yet all similar (as the underlying function is the same for all).

Differential calculus deals more easily with continuous lines and points of inflection than with gaps and angles. Lynn's and Cache's writings of the mid-nineties emphasized the role of mathematics, calculus, and continuous functions as new tools of design,[72] and Lynn's 1996 essay on "Blobs" immediately captured the spirit of the time.[73] The blob itself quickly became a visual and notional trope of the end of the twentieth century. Toward the end of the decade the fling of digital architects with topological geometry further amplified this tendency toward formal continuity. By 1999, from car design to web design, from sex appeal to fashion magazines, curvaceousness was ubiquitous,[74] and from the Guggenheim Bilbao on, curvilinearity was often singled out as the diacritical sign of digital design. The new organicist and morphogenetic theories[75] that crossed paths with the mathematical ones around that time would eventually become staples of digital design theory. Technical factors clearly drove the tilt toward the curve that marked end-of-millennium digital design, but it is the deeper empathy between digital technologies and the more general postmodern and posthistorical aura of the nineties that best explains the spirit of the new culture of digital design that was then taking shape.

The first digital critics of the time were not at leisure to investigate the matter, due to the sudden disappearance of what was then called "the new economy." The dot-com bust of 2000–2002 had both direct and indirect consequences for digital designers,[76] and in the more sober environment that prevailed after 2001, digital theory often inclined toward a more restrained process-conscious approach, to the detriment of its earlier formal glamour. If the continuity between digital design and fabrication tools had been first exploited primarily to produce showcase pieces of unique and sometimes virtuosic formal difficulty, the accent now shifted toward the technical and social implications of a fully integrated design and production chain.

The capacity to mass-produce series of nonidentical items led to a new range of theoretical and practical issues. The idea of nonstandard seriality, as this mode of production is often called,[77] was already inherent in the original definitions of the objectile, but its economic implications were not. In its simplest formulation, the theory of nonstandard seriality posits that economies of scale are irrelevant in digital production processes: every item in a digitally produced series is a one-off. Industrial mass production used to depend on mechanical matrixes, molds, or casts of which the upfront cost had to be amortized by reusing them as many times as possible. But due to the elimination of mechanical matrixes, digital fabrication tools can produce variations at no extra cost, while product standardization, still a perfectly reasonable option in many cases, and still high in demand for a number of reasons, has nevertheless lost its main economic rationale. In a digital production process, standardization is no longer a money-saver. Likewise, customization is no longer a money-waster.

Nonstandard seriality, in turn, already contains the seeds of a potentially different authorial approach. As digital fabrication processes invite endless design variations (within given technical limits), and promise to deliver them at no extra cost, the question inevitably arises as to who is going to design them all. In a parametric design process, some parameters are by definition variable. This variability may be automated and machine-controlled: for example, a program may be instructed to generate any number of variations, randomly or as a function of some external factor. Alternatively, the designer may choose and fix all parameters that determine each individual item right from the start, thus "authorizing" only a given number of them—a closed series of different objects all designed by the same author. But a third possibility cannot be ruled out: some parameters may be chosen, at some point, by someone other than the "original" author, and possibly without his or her consent. Open-endedness and interactivity are inherent in the notion of digital variability, but this participatory approach to digital design has only recently gained wider recognition, in the new technocultural environment of the so-called Web 2.0,[78] and in the context of the current excitement for all forms of collaborative and "social" use of the new media.

Some earlier, pre-Web 2.0 experiments on interactive design formats have recently given way to heavyweight, full-fledged (and heavily advertised) software platforms aimed at design collaboration. Most such tools have been developed to facilitate the flow of technical information among teams of designers working on the same project, but the potential import of this participatory approach is vaster and deeper, and it suggests more imaginative modes of use.[79] Engineers already fret about the dilution of responsibilities that digitally supported collaborative design methods may entail. But what if the same tools were used to involve, at the opposite end of the chain, the patrons or owners, for example, as well as clients, end users, customers, or citizens? What if some parts of the design process itself could be made interactive and public?

Digital technologies inevitably break the indexical chain that, in the mechanical age, linked the matrix to its imprint. Digital photographs are no longer the indexical imprint of light onto a surface; digitally manufactured objects are no longer the indexical imprint of a mold pressed into a metal plate; and digital variability may equally cut loose the indexical link that, under the old authorial paradigm, tied design notations to their material result in an object. In a digitized design and production process, the Albertian cutoff line that used to separate conception and construction is already technically obsolete. But if Alberti's allographic model is phased out, the traditional control of the designer over the object of design (as well as the author's intellectual ownership of the end product) may be on the line, too. If variations may occur at any time in the design and production process, and if parts of the process are allowed to drift open-endedly, interactively, and collaboratively, who will "authorize" what in the end? Interactivity and participation imply, at some point, some form of almost collective decision-making. But the wisdom of the many is often anonymous; anonymity goes counter to authorship, and, since the inception of the Albertian model, authorship has been a precondition for the architect's work.

Yet we can already count plenty of instances where the new digital media are fast unmaking established traditions of authorship that, until a few years ago, would have been deemed indispensable—both intellectually and economically. Who could have anticipated the meteoric rise of a universal encyclopedia that has no author (because it has too many), and which everyone uses (with some precautions) but no one pays for? Open source software is developed in the same way. The music industry has already been upended by the sheer impossibility of enforcing copyright law in the digital domain. The bottom line seems to be that digital technologies are inherently and essentially averse to the authorial model that rose to power with mechanical reproduction, and is now declining with them.

The old authorial paradigm was predicated upon mechanical indexicality, and on the mark of authorship that mechanical reproduction carried over from archetypes to identical copies. The rise of architectural authorship may have been following the same trajectory. If this is the case, then chances are that, with the transition from mechanical to digital technologies, and from identical to variable reproductions, a recast of architectural agency will also be inevitable. In fact, the trend may already have started.

1.7 The Reversal of the Albertian Paradigm

At the beginning of the Modern Age, the power of identical copies arose from two parallel and almost simultaneous developments: on the one hand, identicality was an intellectual and cultural ambition of the Renaissance humanists; on the other, it would soon become the inevitable by-product of mechanical technologies, which it has remained to this day. It is Alberti's precocious and relentless quest for identical copies of all kinds that makes his work so revelatory in this context. Most of his inventions failed, but many of his ideas thrived. Predicated upon the same mandate of identical reproducibility (in this case, the identical translation from project to building),[80] Alberti's definition of architecture as an authorial, allographic, notational art held sway until very recently, and defines many if not all of the architectural principles that the digital turn is now unmaking.

The shaping of complex geometries and of irregular, ungeometrical or "free" forms, which was the first and most visible achievement of the digital turn in architecture, may have been a transient incident. But due to CAD-CAM integration, and counter to the Albertian principle of separation between notation and construction, digital architects today are increasingly designing and making at the same time. Acting almost like prosthetic extensions of the hands of the artisan, digital design and fabrication tools are creating a curiously high-tech analog of preindustrial artisanal practices. Traditional craftsmen, unlike designers, do not send blueprints to factories or building sites: they make with their hands what they have in their minds. The objection, so frequently raised, that this new mode of digital artisanship may apply only to small objects of manufacturing is theoretically irrelevant: any big object can be assembled from smaller, digitally fabricated parts.

Ultimately, Alberti's modern and humanistic authorial tenet, which called for the final notation of an object (its blueprint, in twentieth-century parlance) to be materially executed without any change, may also be doomed in a digital design environment. Projects (and not only for buildings: the principle can be generalized) are increasingly conceived as open-ended, generative scripts that may beget one or more different objects—redesigned, adapted, messed up, and tampered with by a variety of human and technical agents, some of them uncontrollable and unpredictable.

So it will be seen that, over the brief span of less than two decades, the digital turn may have already undermined many of the basic principles that defined modern Western architecture from its Albertian beginnings. In the course of the last five centuries, the power of exactly repeatable, mechanical imprints has gradually shaped a visual environment where identicality is the norm, similarity insignificant, and the cultural expectation of identical copies ultimately affects the functions and value of all signs. Under this semantic regime of modernity, only signs that are visually identical have identical meanings. This is the way modern logos, emblems, and trademarks work. They are recorded and protected by copyright laws, which register an original and all identical copies of it—but leave resemblance and similarity in limbo.

Similarity, imitation, and mimesis are essentially premodern, nonquantifiable notions, and as such are hard to appraise in a modern marketplace, and hard to defend in a modern court of law. Before the age of mechanical copies, more complex cognitive processes conferred steady meanings on variable visual signs. The Senate and People of Rome (SPQR) did not legislate the design of their legions' banners, on which fowls of various shapes easily fulfilled the same symbolic function: in any event, everyone knew that the banner of the Roman legion was meant to be an eagle. In ages of variable copies, the meaning of visual signs does not depend on sameness, but on similarity. This was the case in the West before the rise of print, and this is again the case now, in the vast and growing domain of variable digital media.[81]

As Erwin Panofsky claimed in a celebrated and controversial essay, the apparently random drift of late medieval architectural and decorative visual forms was nevertheless derived from, and inscribed within, a set of fixed normative genera. Panofsky famously interpreted this pattern of "differences within repetition" as an isomorphism between Gothic architecture and Scholastic philosophy, both based on a genus-species relation between stable general categories and variable singular events.[82] Equally famously and controversially, and almost at the same time, Richard Krautheimer examined the conspicuous and at times baffling variances between medieval monuments that were meant to be recognizable copies of the same famous archetypes, to conclude that their semantic function was eminently symbolic (or socially conventional), and unrelated to iconicity (or actual visual resemblance).[83] Both analyses aptly describe the visual environment that is being shaped by contemporary digital media. Each objectile is an exactly transmissible but nonvisual notation: it is a fixed normative genus, which may engender infinitely variable visual species. All singular eventuations of the same algorithmic code will be different from, but similar to, one another. They will expect from their beholder a capacity to read and discern similarities, and to use this ancestral cognitive skill to recognize meaningful patterns in a stream of endlessly variable visual signs.

We have reason to worry about, and possibly lament, the forthcoming demise of traditional architectural authorship. The recent spasms of authorial conceit (always an indelible part of the architect's trade, but recently risen to unprecedented levels) further reinforce the perception of an incipient crisis. Evidently, even among practitioners less inclined to theoretical speculation, the nagging feeling that something today is not quite right with architectural authorship has made some headway. But the likely victim of today's upheavals may not be the general, timeless notion of architectural agency. Once an esoteric modernist theory, now an ordinary postmodern practice, the death of the author affects today but one, particular, time-specific category of authors: the author of identical, mechanical copies—the modern, Albertian author. Modern objects (authorial, authorized, and identically reproduced) might also disappear in the process. But many other modes of agency remain, and the old ones that were in force before the rise of the Albertian paradigm could help anticipate the new ones that may come after it.

The Scholastic flavor of the objectile may be an accidental side effect of its Deleuzian, Leibnizian, and mathematical provenance. The objectile is to an object what a mathematical function (a script or notation) is to a family of curves, or the Aristotelian form is to an Aristotelian event: in Aristotelian terms, the objectile is a generic object. The theory of the objectile also

2

THE RISE

Like all things handmade, all that is digital is variable. Five centuries of mechanization almost made us unmindful of the fact. The transition from manual variability to mechanical sameness is often seen as a unique accomplishment of the industrial revolution, but well before the age of steel and coal, visual standardization was inaugurated by mechanical printing, and by the new ideas of Renaissance humanism. At the end of the Middle Ages, Alberti aimed at identical reproductions of almost everything: of texts and images, of letters and numbers, of drawings and designs, of paintings and sculptures, of architectural parts, occasionally of entire buildings, and of other three-dimensional objects, both natural and man-made; in short, of almost every manifestation of art and nature. Alberti's relentless pursuit of identical copies highlights one of the most significant turning points in the modern histories of art, science, and cultural technologies, and his theory of design as a new way of "building by notation," predicated upon and deriving from the same ideal of identical reproducibility, still defines the architectural profession around the world. Yet identical reproduction was, with few exceptions, technically impossible, as well as culturally irrelevant, at the end of the Middle Ages. In the absence of suitable cultural technologies, Alberti had to invent some new ones; a few of these were successful, but many were not—and were in fact so odd and untimely that they

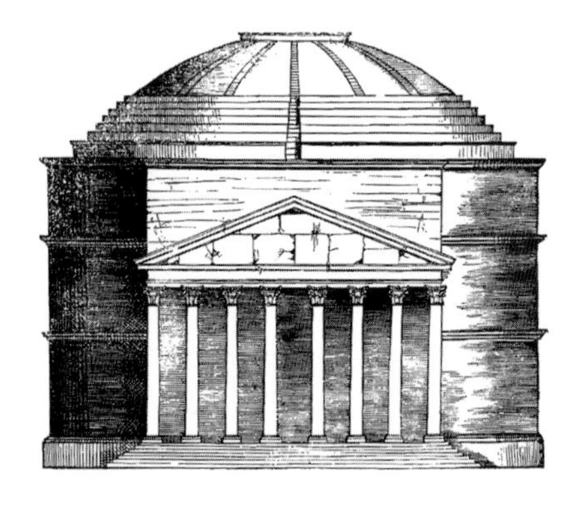

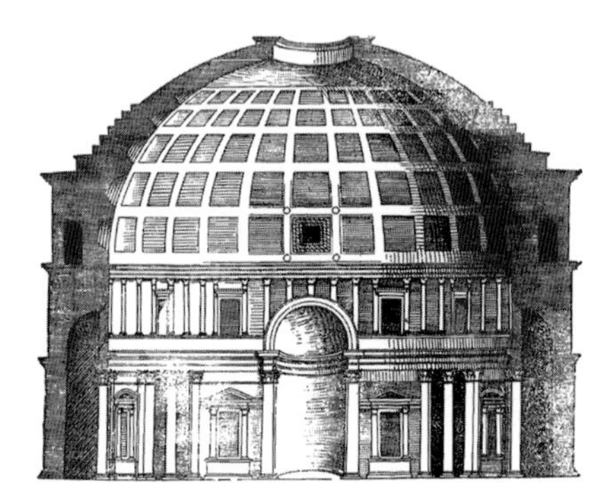

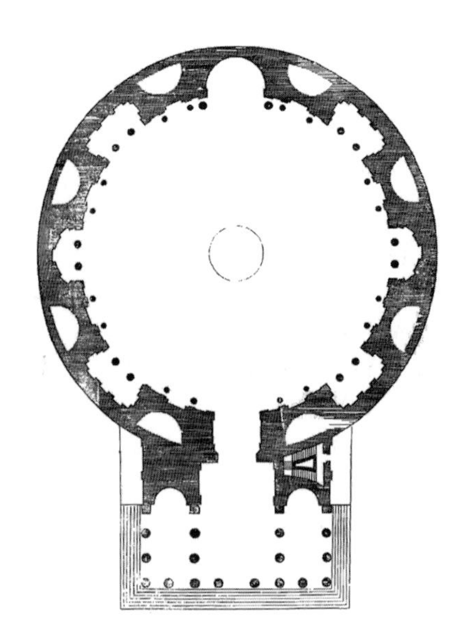

were largely misunderstood or ignored when Alberti first formulated them, and for a long time thereafter.

2.1 Alberti and Identical Copies

Conversant as he was with the sadly corrupted remains of classical literature, Alberti knew all too well that the manuscript transmission of texts and images was a risky operation. Copyists make mistakes, sometimes they interpret, sometimes they interpolate, sometimes they invent. But Alberti also knew, as everyone did since classical antiquity, that the risk of copying affected the transmission of text and images in a markedly asymmetrical way: alphabetical texts, and numbers, were known to travel across space and time faster and more safely than pictures. And, at the time of Alberti, most scholars would have known that the fidelity of a handmade copy of a manuscript drawing was also in inverse proportion to the complexity of the drawing, and to its distance from the archetype. Simple geometrical diagrams, for example, could survive several rounds of copying relatively unscathed, but richer or less geometrically definable drawings could not. Alphabetical texts (and numerical sequences, both in the Hindu-Arabic and in the Latin notations) have one advantage over handmade drawings: letters and numbers, all together, make for a short list of exactly repeatable standardized signs; a drawing is made up of an unpredictable number of signs, none of them standard or exactly repeatable.[1]

2.1

Photomontage. Plan, front view, and section (with perspectival inserts) of the Pantheon from three different pages of the first edition of Sebastiano Serlio's Third Book (*Terzo libro ... nel qual si figurano, e descrivono le antiquità di Roma* [Venice: Marcolini, 1540], pages VII, VIII, IX).

Given this predicament, which he shared with all classical and medieval authors, it is not surprising that Alberti should take steps to preserve the fidelity to the original of the copies of images he valued, used, and needed. More surprising, however, was the extremity of the means to which he resorted. Whenever precise reproduction was of the essence, Alberti simply chose not to use illustrations that would not have been reproducible, replacing them instead with textual descriptions (*ekphrasis*) or other numerical or letter-based stratagems.[2] None of Alberti's three famous treatises on painting, architecture, and sculpture was illustrated, nor ever meant to be.[3] In some cases Alberti resorted to elementary geometrical diagrams, simple enough to be easily reproducible. In other cases, oddly, Alberti tried to counter the failings of analog images by digitizing them, in the etymological sense: replacing pictures with a list of numbers and a set of computational instructions, or algorithms, designed to convert a visual image into a digital file and then recreate a copy of the original picture when needed. The alphabet, the diagram, and the algorithm: each of these cultural technologies could somehow substitute for nontransmissible pictures, but each had its own shortcomings.

2.2 Going Digital

One of the most original results of Alberti's research into image-making technologies is the famous digital map he published in a brief Latin work, the *Descriptio urbis Romae.*[4] At some point, presumably in the late 1430s or 1440s, Alberti had painstakingly surveyed, measured, and accurately drawn to scale a map of the city of Rome. But manual copies in this instance could hardly have preserved the measurements of the original map, and as this drawing could not be adequately translated into words, Alberti found a way to translate it into numbers. Alberti explains in a brief introduction how he had drawn the map, then "digitized" it using a system of polar coordinates. The rest of the book is in fact a list of numbers; the readers were expected to recreate the picture, identical or proportionally identical to the original, by feeding those numbers into a specific instrument, which Alberti also describes, and which today we would call a plotter.[5] Apparently Alberti thought, with some reason, that his high-definition picture would better travel in space and time encapsulated in a digital file, and that the original map should be redrawn each time anew based exclusively on the numerical data—rerunning the program, so to speak, and not via the "analog," manual copy of other preexisting copies of it. Some thirteen centuries earlier, the rationale for this process had been lucidly argued in Ptolemy's *Geography*, or *Cosmography*—the Latin translation of which Alberti knew well.

Alberti went beyond his likely mentor, however, when he applied the same principle to three-dimensional objects: his treatise on sculpture, *De statua*, includes the lengthy technical description of another improbable machine, this one meant to scan human bodies and translate them into lists of three-dimensional coordinates. As in the *Descriptio urbis Romae*, the key piece of hardware in Alberti's *De statua* was a revolving instrument, a wheel of sorts, in this case somehow inconveniently nailed to the head of the body to be scanned. After all salient points had been digitized, the resulting list of numbers would have enabled the original body to be reconstituted and replicated *ad infinitum*, in distant places and at future times, at the same scale or proportionally enlarged or reduced. Alberti also suggested that, using the same technique, different parts of the same statue could be manufactured simultaneously in different workshops: the heads in Tuscany, the feet in Greece, and so on, and when the different parts were assembled, they would all fit together perfectly.[6]

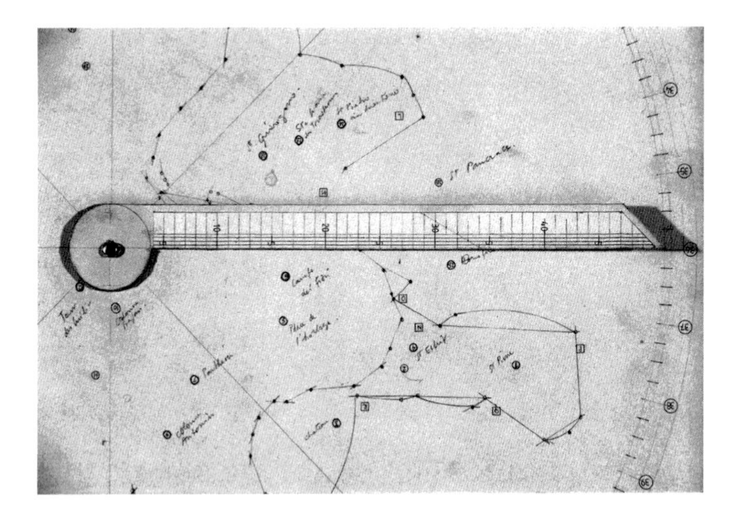

2.2 and 2.3
Reconstruction of the map and drawing device following the methods and coordinates supplied in Alberti's *Descriptio urbis Romae.* Images by and courtesy of Bruno Queysanne and Patrick Thépot.

Alberti's notion that the perpetuity of a monument would be guaranteed by a sequence of numbers better than by the original monument itself may sound odd. Daily experience suggests that stone and marble are stronger and more resistant to time than parchment or paper. And the idea that a statue can be made through the assembly of prefabricated parts does not fail to evoke for us an uneasy feeling of technological *déjà vu*. Why should Alberti dream of, envisage, or prefigure the protoindustrial production system, Taylorism, the assembly line, outsourcing, and computer-aided manufacturing? The makers of today's "file-to-factory," CAD-CAM technologies tout many of the same advantages that Alberti had mentioned. And in another instance of the kinship between premechanical and digital technologies, Alberti's "do-it-yourself" map of Rome curiously resonates with the logic and functions of today's digital mapping and geographic information systems.

For centuries, even after the diffusion of printed maps, specialty maps (particularly for the military, or travel itineraries) were still hand-drawn and custom-made on demand, to include all and only the information required by users. This was already the mode of use that Ptolemy had imagined for his *Geography*, which was originally conceived as an open geographical database—a customizable mapmaking kit of sorts.[7] Likewise, today's web-based digital map services for the general public (offered for free, but with advertising included) generate special scalable maps and itinerary diagrams at each customer's request, and each map includes only the information that the customer (or the advertisers) deem relevant to its purpose, location, and scale. Similarly, geographic information systems for professional use collect information from databases of all kinds to draw, at each request, a new map or diagram, where layers of visualized information can be added at will. The age of the all-purpose geographical printed map (published by various and often national cartographic services at different standard scales for different classes of users, from surveyors to tourists to the military) started relatively late in time, and is already over. The only significant difference between Alberti's digital processes and ours is speed. This is probably one reason why—until very recent times—no one used or even tried to make sense of Alberti's digital tools. Other equally ingenious image-making devices of Alberti's were more successful.

2.3 Windows

In the first book of *De pictura*, Alberti famously defined painting as the result of the intersection of visual rays (or of a visual pyramid, the vertex of which is the observer's eye) with the surface of painting, and on that basis he outlined the principles of the geometrical construction that would eventually be known as central perspective. In the second book of the same treatise, however, Alberti suggested other ways to obtain the same results, all based on the physical, real-size reconstruction of the three-dimensional process of viewing. The procedure is well known. First, the eye of the observer is located at a mechanically fixed point. Then, somewhere between the eye and the object of vision, the picture plane—which Alberti had described as "an open window,"[8] "transparent, and like glass"[9]—is replaced by a solid but translucent surface, a "veil,"[10] which intercepts all visual rays connecting the eye and the object of vision, and where all points of intersection can be physically pinned down and marked or plotted. This creates on the veil a matrix of the picture, which must eventually be copied at the same scale, or proportionally enlarged or reduced, to be transferred onto the actual surface of painting. For this purpose Alberti devised two slightly different methods, based on the use of parallel lines or of an orthogonal grid. In the Latin version of *De pictura* Alberti adds that this procedure was his own invention.[11] We do not know what Brunelleschi (to whom Alberti

2.4
Albrecht Dürer, *Vnderweysuug* [sic] *der Messung, mit dem Zirckel vnd Richtscheyt, in Linien Ebnen vn[d] gantzen Corporen, durch Albrecht Dürer zusamen gezogen, vn[d] durch jn selbs (als er noch auff Erden war) an vil Orten gebessert, in sonderheyt mit xxii Figure[n] gemert, die selbigen auch mit eygner Handt auffgerissen* (Nuremberg: H. Formschneyder [i.e., Hieronymum Andreae], 1538), fig. 67.

dedicated the Italian version of the treatise)[12] would have made of this statement, which most likely refers to the perspectival machine as a whole, not to the grid drawn on the "veil," which had been used from time immemorial for the manual copying and scaling of drawings.

Alberti does not suggest frequencies for the horizontal or vertical lines of the grid, hence the graphic resolution of his window has no preset standards. But if the resolution of the grid is pushed to the limit, its squares become points, or, as Friedrich Kittler has recently suggested, pixels.[13] Following the example of Ptolemy, Alberti might have rasterized his picture using a simple combination of longitude and latitude, whereas today we would proceed by sampling and quantization, but either way, the end product would be the same: a number-based matrix, or, indeed, a digital file. A system of orthogonal coordinates superimposed upon, or underlying, a picture points to quantification and measurability: measurements mean numbers, and Alberti's "grid" method—albeit *prima facie* analog and geometrical—implies the possibility of translating a full image on a screen into a frame of lines and dots, and of recording the position and value of each of these points as a set of numbers.[14]

Alberti's pictorial images are geometrically defined as the imprint left by a cone of visual rays on a surface—the perspectival picture plane. Today we would call this indexical trace, etymologically, a photograph: an image that is automatically drawn by light. In a sense, both of Alberti's perspectival methods—the optical machines from the second book of *De pictura*, but also the geometrical construction from the first book—are designed to capture a snapshot of the images that would ideally take shape on the surface of Alberti's picture plane, "veil," or "window."[15] Once again, Alberti's numbers, letters, diagrams, and algorithms are laborious and often inadequate substitutes for the real protagonist *in absentia* of his image theory: the modern, exactly repeatable, mechanically made, and imprint-based visual copy. Here, as elsewhere, Alberti's tools and processes were intended to curtail, and ideally to eliminate, the variability of hand-making from the process of making copies. Albeit confined to a specific technological ambit, Alberti's emphatic aversion to the role and relevance of the human factor in most arts and crafts may seem at odds with the natural vocation of a humanist—or at least with today's current meaning of the term. This is but one of the paradoxes that mark Alberti's quest for handmade standardization and for individually crafted, custom-made, but perfectly identical reproductions.

In the ninth book of *De re aedificatoria*, Alberti outlines the general principles of beauty as dictated by his trademarked definition of *concinnitas*: "the absolute and fundamental rule in nature." Alberti's *concinnitas* in turn arises from the rightful composition of three qualities: *numerus*, *finitio*, and *collocatio*. While the first and second terms of the triad refer to theories of proportions, and have been lavishly commented upon, the third has often escaped scrutiny. Most recent translators have preserved some of the opacity of Alberti's original, but Jean Martin, who around 1550 first translated *collocatio* into French as *égalité*, or sameness, had been more perspicacious. Alberti requires all parts of a building that are visually related by way of alignment, symmetry, or proximity, to be identical, and he concludes that he always admired the capacity of the ancients to reproduce statues and sculptures that "were so similar to one another that we might claim that here nature herself has been surpassed, since never in her works do we see so much as two identical noses."[16]

2.4 ID Pictures and the Power of Facsimiles[17]

By coincidence, when Alberti, relatively late in life, started a new career as an architect, one of his first assignments was to design a look-alike—the identical replication of an existing building. Alberti's patron, Giovanni Rucellai, wanted a copy of the shrine of the Holy Sepulcher in Jerusalem to be rebuilt in his family chapel in the church of San Pancrazio in Florence. Rucellai's plans were already documented in 1448, and the idea that the new shrine should be a replica is confirmed by the inscription still visible above the entry door ("sacellum ad instar iherosolimitani sepulchri"),[18] dated 1467.

But the copy that Alberti built in Florence does not look like the original, as it stood in Jerusalem around that time. Only a few quirky details are similar. Current criticism assumes that Alberti would have striven to improve upon the model, or to evoke it as it might have existed in some pristine state, implying that Alberti, a great artist, would certainly not have been satisfied with the mere knockoff of a famous building. In point of fact, the knockoff of that unseen and inaccessible building would have been—in Florence, at that time—a challenging and perhaps impossible task. Alberti could hardly have been inspired by a visual model that he and his client had never seen, that no one could see, and on which no reliable visual documentation existed or could be obtained.

A more detailed story of the Rucellai shrine would reveal some odd twists and turns, and require a lengthier review.[19] However, those similarities that exist between the original and the copy suggest that Alberti might have derived his design from textual sources of some kind, received in writing or orally, including some measurements of the original building, but in the absence of any visual documentation. This mode of working was common in the Middle Ages, when words traveled faster and farther than pictures; but in this specific instance, Alberti's predicament once again emphasizes—and emblematically materializes in a building—the conflict between his new scopocentric ambitions and the traditionally logocentric tools at his disposal. Evidently, the pursuit of identicality between a copy and an invisible original is *per se* a paradox: no wonder the original and the copy do not resemble each other. How could they? But also: Why should they? Who would have noticed? Who could have told the difference, and who would have cared at the time?[20]

Many more instances of the central role of identicality in Alberti's thinking, art, and architectural theory and design might be added. Alberti was himself a painter and sculptor, although his talents as a figurative artist were not unanimously praised at the time.[21] Among a handful of Alberti's paintings mentioned in Vasari's *Vite* is a self-portrait that Alberti would have made with the help of a mirror,[22] and self-portraits, together with portraits of friends, are the only paintings cited in Alberti's anonymous *Life* (or autobiography): Alberti painted his own head and portrait, says the author, so that from these images he could be more readily recognized by visitors who did not know him; Alberti also tested the realism of painted portraits by asking children to recognize who was represented, and if the test failed, "he denied to such works the status of art."[23]

At the end of *De pictura*, Alberti asked his readers to reward his authorial efforts by including his own portrait in their future paintings.[24] Artists at the time must have found this request unusual and difficult to comply with. Both the notions of what would be called today an intellectual property to be claimed, and of some royalty to be paid, must have appeared close to meaningless at the time. One may also wonder how Alberti's likeness would have been known to most artists and recognized by the general public. Fortunately, Alberti also bequeathed one or more self-portraits in the form of bas-reliefs on bronze plaquettes—the

2.5 and 2.6
Shrine of the Holy Sepulcher, Cappella Rucellai, Church of San Pancrazio, Florence. Images by and courtesy of Anke Naujokat.

YHESVM QVERITIS N

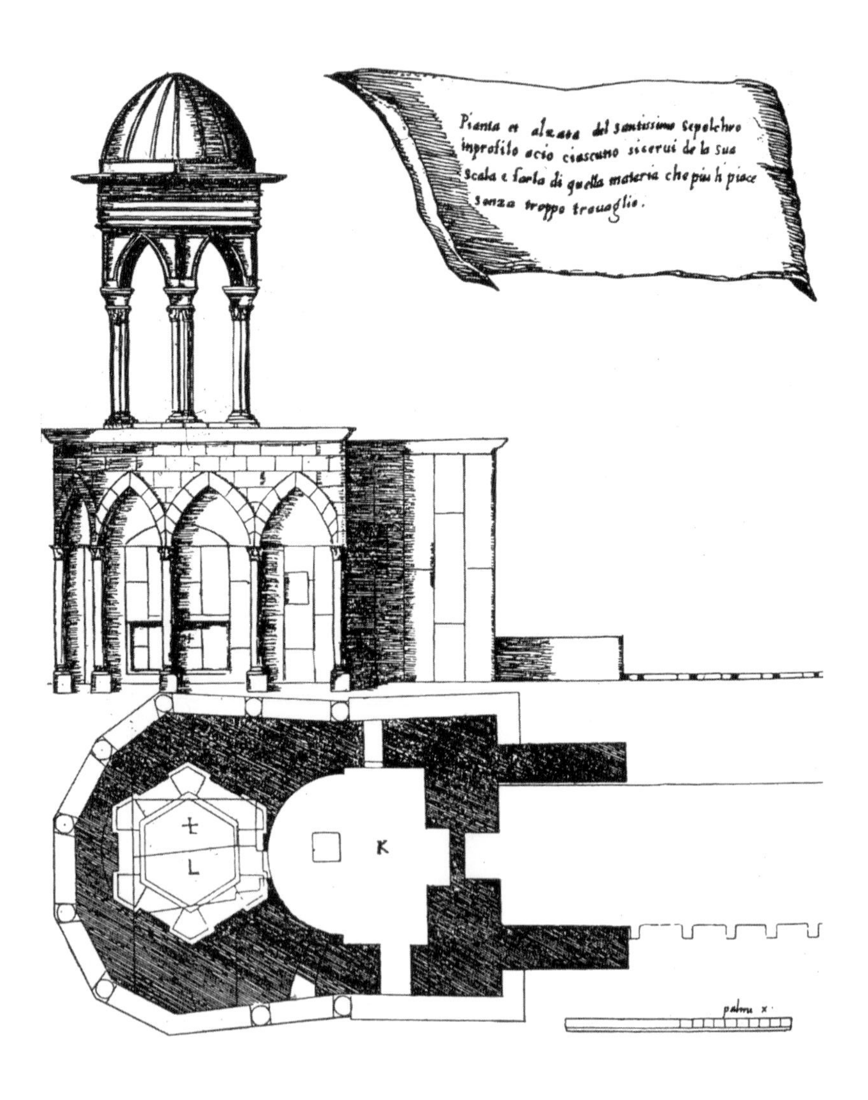
Pianta et alzata del santissimo sepolchro
improfilo acio ciascuno sicerui de la sua
scala e farla di quella materia che piu li piace
senza troppo trauaglio.
K
L
palmi x

2.7
Bernardino Amico da Gallipoli, *Trattato delle piante ed immagini dei sacri edificii di Terrasanta disegnate in Gierusalemme secondo le regole della prospettiva, e vera misura della lor grandezza* (Florence: Pietro Cecconcelli, 1620), plate 33.

closest he got to mechanical reproduction technologies—which today's scholars can still use to scout for portraits of Alberti that may lurk undiscovered in the whole corpus of Quattrocento painting. Alberti's interest in portraiture, and particularly in self-portraits, was once again a matter of identicality. The best painting could only hope to rival a reflection in a mirror or in a still body of water, as both surfaces portray an indexical imprint of light.

2.5 Alberti's Imitation Game and Its Technological Failure

In all of the cases presented so far, Alberti's relentless pursuit of identicality was hindered by technological inadequacy. The same pattern of failure recurred at different points in his work. In the *Descriptio urbis Romae*, *De statua*, and *De pictura*, images that were geometrically defined as identical, or proportionally identical, to their archetypes were obtained directly from indexical traces (the imprint on a picture plane of visual rays converging to the eye, or the imprint of the buildings of Rome on the ground), or through alternative geometrical or mensural processes meant to deliver the same results. Such indexical images, sometimes transferred onto other and more suitable supports, had to be identically recorded and transmitted across space and time. Lastly, copies (whether objects or images) could be recreated identically, or proportional identically to their archetypes—at any place and time and in the absence of the originals themselves. The same logic applies to Alberti's theory of architectural design, but in reverse order: Alberti's project drawings are not indexical images, but are conceived as matrixes in an ideally indexical process.

In all such instances, Alberti's images were meant to be carriers of precise quantitative information, and to record measurable data—data that could be used and acted upon. But this precision,

nologies in Alberti's time could deliver—identical reproductions. Customizable proportional scaling was lost in the process, but evidently the gains more than offset that loss. And mechanical technologies kept producing exactly repeatable visual imprints of all kinds and sizes for the five centuries that followed: a process that was extraordinarily accelerated by the industrial revolution in the nineteenth century, and is only now coming to an end.

Yet, in spite of their untimeliness and quixotic inventiveness, some of Alberti's technocultural inventions famously thrived and, constantly revised and upgraded for the last five centuries, they have marked the history of Western art, architecture, and civilization. Alberti's optical definition of a new class of "photographic" images would not come to full fruition until the development of chemomechanical photography in the nineteenth century, but his geometrical method for determining the intersections of the visual rays with the picture plane was the foundation of modern perspective, and is at the basis of central projections as they are still studied today (or were until recently). Likewise, Alberti's rasterized window inaugurated a long tradition of drawing machines (perspectographs, pantographs, camera obscura and camera lucida devices, etc.) for professional draftsmen and amateurs alike—a sometimes shadowy tradition of craftsmen's tricks and tools of the trade, which art historians and historians of technology have only recently begun to investigate.[26]

More momentously for the history of architecture, Alberti's pursuit of identical replications is also inherently intertwined with his invention of the modern design process. In Alberti's theory a building is the work of the architect that designed it (but did not build it) only if the building and its design can be seen as notationally identical. As mentioned above, Alberti's design process depended entirely on cultural mediators, such as scaled drawings, measurements, and projections, and their respective mathematical underpinnings, and did not call for any mechanical technology, nor require any new instrument or machinery. There is some evidence that in his exclusive reliance on recorded and transmissible project notations Alberti may have once again misjudged the performance of the tools at his disposal, as well as the intensity of the social resistance to his new authorial (and authoritarian) mode of design. Nonetheless, albeit initially resisted by some, Alberti's notational method was eventually adopted by most. For the last five centuries, the centrality of the Albertian notational paradigm to the theory and practice of Western architecture has proven that Alberti's separation of design and construction was viable, and could be technically and socially enforced—within limits.

The constraints that Albertian notationality imposed upon architectural design were a determining factor throughout the history of modern architecture—and have been equally relevant to the recent history of digital design. But in the mid-fifteenth century Alberti's new way of building by notation offered a modern solution to a crucial late medieval problem, which Brunelleschi's achievements and failures as an artisan master builder had famously brought to the forefront only a few years before.

2.6 The Invention of the Albertian Paradigm

Nelson Goodman famously stated that arts that become allographic win "their emancipation not by proclamation but by notation."[27] Had Goodman been more interested in early modern history, he would easily have found the starting point of architecture's "emancipation" from its ancestral autographic roots. The modern history of architecture as an authorial art began with a building, Brunelleschi's dome for the cathedral of Florence; but the new definition of architecture's allographic and notational status came into being only with Alberti's theory and his treatise, *De re aedificatoria*. The building preceded the book, albeit briefly; Alberti's theory took its cue from Brunelleschi's practice, but went far beyond it.

In the somewhat dramatized narrative of his contemporaries (or almost contemporaries), Brunelleschi was the real revolutionary of the two—and, from today's vantage point, the *révolté* avant-gardist and Promethean, proto-romantic hero. The traditional organization of all major, late medieval civic building programs was the main impediment to Brunelleschi's building of the dome at Florence. Brunelleschi managed to build the dome precisely because he could get rid of an old system and impose a new one. His legendary struggle against many personal and institutional foes is vividly recounted by his first Florentine biographers, Manetti and Vasari. The decision-making process at all stages in the construction of the dome, which is well documented, was something similar to what today we would call design by committee, except that most of those decisions would not qualify as acts of "design" in the modern sense of the term. Various committees of officials of the Guild of Woolmakers (Arte della Lana), officials of the Works of the Cathedral (the Opera del Duomo, which exists to this day), and representatives of the master workers kept meeting throughout the construction of the dome, asking for internal and external reports, smashing old models, demanding new ones, deciding on assignments, delegating responsibilities, and making payments.[28] Likewise, from the start, three master builders were appointed to construct the dome, and if Brunelleschi was indeed the first among them, he must have been first among equals.[29] The other prominent master builder on the team was Brunelleschi's old rival, Ghiberti; and in spite of Manetti's claim that Brunelleschi managed (with a stratagem) to have Ghiberti fired at some point between 1423 and 1426, documents prove that Ghiberti was almost right in claiming that he stayed on the job till the end, that is, until 1436.[30] Similarly, the incorporated master workers must have been a vocal presence on the building site throughout, as proven by the story of their strike and of their overnight replacement by some nonunionized craftsmen, whom Brunelleschi apparently called in from Lombardy. Neither the story of the strike nor that of the Lombard strikebreakers is reliably documented, but apparently Manetti, and particularly Vasari, wanted to emphasize that, if Brunelleschi could train a whole new team of inexperienced—and, even worse, Lombard—craftsmen in just "one day," and to his satisfaction, the traditional know-how of the unionized Florentine craftsmen could not have been a determinant technical factor in the building of the dome.[31]

The tales handed down by the chroniclers of the time seem to suggest that, in the perception of his contemporaries, Brunelleschi's main task in getting the dome built was finding endless and ingenious ways of cheating dumb and gullible committees of patrons, colleagues, and subordinates. That may well have been so, but only incidentally. Regardless of the truth, or accuracy, of the anecdotes on record, Brunelleschi was evidently fighting an uphill battle to give shape to the new identity of the intellectual author of a building—in a context where this authority was traditionally vague or ill defined.

Brunelleschi came to the competition for the dome with no experience as a builder, but with a remarkable reputation as a maker—and unmaker—of identities. He was celebrated as the inventor of the then famous prank of the fat woodworker, whereby Brunelleschi had persuaded a Florentine craftsman chosen almost at random that he—the fat woodworker—was not himself, but someone else. The trick worked so well that at the end the victim concluded that he could not longer tell who he was, and that there was no way to find out. Deprived of his Florentine identity, the woodworker then emigrated to Hungary, which was probably almost as far as one could go, in the Florentine perception of distances, prior to the discovery of America. There he started a new life and became rich.[32] In spite of the happy ending, the story is a reminder of the shifting and labile nature of personal identities in early modern Europe—Natalie Zemon Davis's story of Martin Guerre offers a similar case in point.[33] Even more than it affected personal identities, this indeterminacy pertained to all kinds of objects and media objects—from texts to images to music to buildings—whose material instantiations drifted in time and space, alongside the vague identities of their multiple and sometimes anonymous makers. The woodworker's prank proved that even personal and professional identities in Florence were determined by collective social consensus, not by autography. When he set out to build the dome, Brunelleschi decided that should change.

Regardless of the still participatory and collective process of decision-making that accompanied all stages of the building of the dome, Brunelleschi evidently wanted the end result to be seen as his own. One means to this end would have been for Brunelleschi to build the dome by himself—not figuratively but literally, to the last brick. Failing that, the method he devised was the closest possible approximation to full autography—and, as it seems, it worked. Evidently, it took more than one man to build the dome, and workers had to be told what to do. This is indeed one reason, in Goodman's scheme, why some arts become allographic: authors need notations so that others can help them make what they could not do alone. Brunelleschi had an additional concern: he needed to instruct his workers yet keep the authorial role for himself. Today we would not see that as a problem, but Brunelleschi was fighting to create and occupy a new authorial territory, and as in all battles, he had to strategize and take precautions. First and foremost, all his ideas had to stay with him as long as possible: directions had to be given step by step, and strictly limited to what was necessary at any given point in time.

The legend of the models that Brunelleschi would have left deliberately incomplete thus makes perfect sense.[34] See, for example, the famous building program of 1420: after pages of exceedingly detailed instructions and measurements, the program cuts short when it comes to explaining the high point of the project—how to build the dome without centering. On that, the writer (or writers) perfidiously suggests: "only practical experience will teach which is to be followed."[35] Taccola, the Sienese engineer, recorded in writing Brunelleschi's advice not to unveil the inventor's ideas, lest they be stolen by others.[36] Anecdotes abound, from Brunelleschi's models made from carved turnips (turnips are not a durable material, nor one that would sustain elaborate detailing) to his personal inspection of each brick and stone, and even of the sand and lime for making the mortar: as Manetti concludes, "he seemed to be the master of everything."[37] Perhaps the most famous story comes from Vasari, who recounts how Brunelleschi gently crushed an egg on a table to make it stand up (an anecdote also attributed to Columbus). Brunelleschi's rejoinder to the assembly of master builders for whom he staged the demonstration (and who claimed, predictably, that anyone could have done it) is telling: this, he said, is why I'm not unveiling my project to you; if I did, you would build it without me.[38] This is, of course, the very reason why architects still make project drawings: so that others can build in the designer's absence. But Brunelleschi—precisely because his authorial role was not otherwise recognized—needed to make his authorship of the building physically manifest, tangible, and visible at all times. Which he did by being there all the time, distilling instructions only when needed; in Manetti's narration, inspecting and touching and almost dedicating every brick that was laid. Brunelleschi may have invented the modern identity of the architectural author, but his was authorship by indexical imprint—it was autographic, artisanal authorship.

Brunelleschi paid a heavy price for his authorial ambitions. The unfinished models and drawings that he produced for some of his other buildings were occasionally completed by others, who filled in the blanks at their discretion—with Brunelleschi there, but apparently busy elsewhere—and thus made "mistakes."[39] When Brunelleschi saw the errors he was angry and blamed the builders, but he did not raze the buildings or even have the errors rectified. These misadventures may have taught him something about the advantages of notations.

The building of the lantern of the dome started in 1446, only a few weeks before Brunelleschi's death. According to Vasari, Brunelleschi, who had won the competition for the lantern in 1436, wrote in his will that the lantern should be built precisely as he had conceived it (following his "model" and his "indications in writing"); this would suggest that Brunelleschi, *in articulo mortis*, finally gave in and adopted a less incomplete form of architectural notation. Not surprisingly, Manetti tells a completely different story.[40] In any event, the shift toward a new allographic, or notational, definition of the architect's work was not to be of Brunelleschi's making. Architecture's notational turn was being honed and conceptualized, exactly at that time, in Alberti's treatise on building, which Alberti might have finished writing around 1452.

Yet when Alberti, at almost the same time as the writing of *De re aedificatoria*, started an actual career as an architect, he soon found out, to his detriment, that master builders tended to be an independently minded lot. Alberti's insistent and somewhat peevish expostulations in his correspondence with the site manager in Rimini prove that local workers were loath to comply with Alberti's designs, which Alberti had supplied via drawings and models, and further explained in writing.[41]

Besides tradition, mistrust, and ill will, communication itself might have been a stumbling block. Alberti could well ship batches of scaled drawings of plans, elevations, and side views, as per his own prescriptions in the second book of *De re aedificatoria*; but he could hardly expect that local workers in a remote, backwater location could read and understand project drawings in the absence of the designer himself—there to explain, gesticulate, rage and rail on site, as Brunelleschi did throughout the building of the dome. Alberti's vision of a standardized notational language for the exchange of three-dimensional technical data would take some centuries to prevail, and around 1450 his new notational way of building could have been deemed futuristic, if the notion had existed at the time. Not surprisingly, some of Alberti's contemporaries made their reservations known. Manetti's (and later Vasari's) story of Brunelleschi's heroic deeds should be seen in the light of Manetti's stance against Alberti's method of design, which Manetti insinuated was presumptuous and unworkable.[42] Manetti may have written his *Life of Brunelleschi* in the immediate aftermath of the *editio princeps* of Alberti's *De re aedificatoria*, and it is known that Manetti ended his life as a repentant Savonarolan—a Piagnone.[43] Vasari too bore Alberti little affection, and he might have cherished the Brunelleschian myth of a Florentine artisanal humanism, prior to and immune from Alberti's megalomaniac, cosmopolitan intellectual ambitions.

Even today, and in spite of the fact that Alberti's authorial and notational method of design has become a staple of modern life, some still find it a bit too high-tech for their taste. After all, Alberti posits the complete disembodiment of the process of making objects. The Albertian object is but the mechanical reification of an authorial script, and in Alberti's theory the material process of making, albeit carried out by human hands, is devoid of all human intentions. John Ruskin famously said that Renaissance architecture turned workmen into slaves;[44] Karl Marx might have more pertinently remarked that Alberti's method would ultimately "alienate" artisans and turn them into proletarians. Contemporary Heideggerians might add that Alberti's method started to turn objects from living *Ding* into muted *Gegenstand*, implying that Alberti was up to no good.

Regardless of what Alberti was really up to, it did not happen for quite some time. When, early in the second half of the twentieth century, Nelson Goodman assessed the notational status of the modern blueprint, he concluded that even then architecture's conversion to a full allographic mode was not yet completed. Architecture appeared to him to be still in a "mixed" and "transitional" state—halfway between its autographic past and some foreseeable allographic future. As Goodman wrote, he was not entirely "comfortable about identifying an architectural work with a design, rather than a building," and most architects at the time would have agreed.[45]

That very transition is now several years in the making. Contemporary CAD-CAM technologies have simply obliterated the notational gap that for centuries kept design and construction apart. Each CAD file contains the precise and univocal denotation of the position in space of each geometrical point that composes a building, and a digital notation can be executed anywhere, anytime, regardless of the presence or absence of its author, so long as a machine similar to the one used to make that file is available to read it.[46] A CAD file would certainly satisfy all of the complex requirements that Goodman asked of notational systems. With CAD-CAM technologies, architecture may have finally attained full allographic status.

Digitally generated, automatic allography, however, may also bring unexpected consequences. Owing to CAD-CAM integration and BIM (building information modeling) software, design and production will increasingly merge and overlap in a single, seamless process of creation and production. In some cases, limited for now to the fabrication of small objects and prototyping, existing CAD-CAM technologies have already achieved that stage: an architect's design can be immediately and automatically fabricated—if need be, in front of the architect and while the architect is still working on it. This is not made by magic—nor by slaves, as Ruskin claimed. It is done by a digital CAD-CAM machine. As advanced CAD-CAM systems already support and, indeed, encourage cooperation and interaction among human actors and technical networks in all stages of design and production, the end result of full, digitally supported notationality in architecture may also reenact some of the original, ancestral, and autographic aspects of artisanal hand-making. One can discuss, design, and make at the same time—just as premodern artisans and pre-Albertian master builders once did. Perhaps the late Heideggerians of our time should have a second look at digital technologies.

3

THE FALL

Since the rise of Renaissance humanism and the beginning of the Early Modern Age, three instances of identical reproducibility have marked Western architectural history: the identical translation of design *notations* into physical buildings; the identical transmission of architectural *information* through space and time; and identical *fabrication*, or the pursuit of economies of scale through mass production and standardization. Of these, the latter grew exponentially during the nineteenth and twentieth centuries, due to the industrial revolution; but from the viewpoint of an axiology of identical copies, industrial mass production is little more than an extension and amplification of the technical and cultural trend that started with printing and, in architecture, with what I have called "the Albertian paradigm." As John Ruskin remarked, print was the original sin and the germ of all industrial evils to follow, because it is "that abominable art of printing" that "makes people used to have everything the same shape."[1]

As we now know, Ruskinian Gothic did not in the end stop the industrial revolution. At the apex of the industrial age, sometime during the second half of the twentieth century, mechanical implements and a number of related cultural technologies brought to completion an ideal state of almost seamless continuity in the identical reproduction of all objects of design and manufacturing. The modern blueprint format (and its Mongeian geometrical

3.1
Greg Lynn, Alessi Coffee & Tea Piazza.
Copyright Greg Lynn FORM, 2000 ©.
Courtesy of Greg Lynn FORM.

suggested another suitable dichotomy. In an essay first published in 1929, the psychologist Wolfgang Köhler compared two abstract visual shapes, one rounded and one angular, and two made-up names, "takete" and "maluma." When asked, most subjects associated the name "takete" with the angular shape, and "maluma" with the round one—apparently, regardless of language, culture, or environment.[3] This famous experiment was eventually used to corroborate various controversial theories. There is no need to delve into such risky cognitive matters, however, to assume that a similar opposition between angular and round forms may have equally played out in the history of Western architecture. The Bauhaus building in Dessau is taketian, the Einstein tower is malumian; the Seagram Building is taketian, the former TWA terminal at JFK Airport in New York City is malumian.

During the last decade of the twentieth century, architecture was conspicuously malumian. One should not conclude from this that all *fin-de-siècle* architecture has a tendency toward rotundity. This analogy, which holds good for the end of the nineteenth and of the twentieth centuries, might not be sustained over a longer time span. But toward the end of the nineties, malumianism was almost aggressively ubiquitous. It dominated industrial design, fashion, furniture, body culture, car design, food, critical theory in the visual arts, sex appeal, the art of discourse, and military engineering—as well as architecture.[4] Admittedly, one of the most influential architectural writers of the decade, Rem Koolhaas, kept designing in a taketian mode, but the most iconic building of the time, Gehry's Bilbao, was malumian. Malumian folds and blobs were theorized, advocated, and patronized, and in the late nineties this malumian tendency was frequently associated with topological geometry and with the use of digital technologies. Indeed, "topological" architecture, as it was then called,[5] was seen for a while as the quintessential embodiment of the new computer age—and we all remember the excitement and exuberance that surrounded all that was digital between 1996 and 2001.

The notion of a cause-effect relationship between digital technologies and free form, or complex geometries (including the most complex of all, topology), was built on a truism but generalized into a fallacy. True, without computers some of those complex forms—most of them round—could not have been designed, measured, and built. Yet digital tools *per se* do not necessarily impose malumian shapes. In fact, toward the end of the nineties, computers were routinely used by many unimaginative architects to design perfectly banal and unremarkable buildings, where the use of digital tools did not leave any visible trace. Why, then, did the theoretically savvy digital avant-garde of the nineties adopt and embrace, with almost unanimous enthusiasm, folds and blobs, rotundity and seamlessness, smoothness and continuity, and make of these purely visual tropes the diacritical sign of a new technical age? Only a dialectic relationship between technology and society can bring about enduring technosocial transformations—and with them, meaningful changes in architectural form. As a new course in architectural design, and in form-making processes, started to take shape in the late nineties, it is legitimate to wonder how, when, and why this happened, and to what extent the arguments that triggered these changes may have stood the test of time and still hold true to this day.

At the beginning of the nineties, architectural theory was still busy discussing deconstructivism and its eminently taketian avatars in building. American critical theory of the time was under the influence of a few Parisian thinkers—often ignored in their homeland. When Gilles Deleuze's impenetrable book on *The Fold: Leibniz and the Baroque* was first published in France in 1988, it failed to generate much critical acclaim in the immediate surroundings of Boulevard Raspail.[6] Yet the Deleuzian fold was granted a second lease on life when Peter Eisenman—starting with the first publications of his Rebstock project in 1991—began to elaborate an architectural version of it.[7] This appropriation was secured in 1993 by a special issue of *Architectural Design*, suitably titled "Folding in Architecture,"[8] which was edited and prefaced by Greg Lynn, a former student of Eisenman's who had worked as a project assistant on the Rebstock proposal. At the same time, Deleuze's book was published in English (and a chapter of the translation was included in the same issue of *Architectural Design*).[9]

Deleuze's book was on Leibniz, on folds, on the baroque, and on many other things as well. Most of it can be read as a vast hermeneutic of continuity, which Deleuze applied to Leibniz's theory of ideas (including his notorious monadology), to Leibniz's mathematics (differential calculus in particular), and to various expressions of the baroque in the arts: the fold, a unifying figure in which different segments and planes are joined and merge in continuous lines and volumes, is both the emblem and the object of Deleuze's discourse. Had Deleuze been more interested in the history of art, he would have found a perfect incarnation of his Leibnizian fold in Hogarth's only slightly later "line of beauty," or serpentine line, which, just like Deleuze's fold, can be described as a mathematical point of inflection in a continuous line.[10] Hogarth did not resort to differential calculus to define his ogee-shaped serpentine—nor did Deleuze, for that matter. Deleuze always referred to the fold in visual, geometrical terms (as the point of inflection that separates concavity and convexity in a curved line, or the point where a tangent intersects a curve), not in the abstract terms of modern differential calculus (where the point of inflection is defined as the maximum or the minimum in the first derivative of the function of the curve).[11] In Deleuze's reading of Leibniz, the fold more generally epitomized the spirit—not the technology—of a new mathematics of continuity: folds avoid fractures, overlay gaps, interpolate.

Eisenman's reading of Deleuze's fold in turn retained and emphasized the notion of forms that can change, morph, and move, and of a new category of objects defined not by what they are, but by the way they change and by the laws that describe their continuous variations. Eisenman's essays prior to 1993 also reveal a significant topical shift, evolving from a close, often literal interpretation of Deleuze's arguments (in 1991 Eisenman even borrowed Deleuze's notion of the "objectile")[12] to more architecturally inclined adaptations, including the use of René Thom's diagrams as design devices for generating architectural folds—a short circuit of sorts, as Thom's topological diagrams are themselves folds, and Thom actually itemized several categories of folding surfaces.[13] In his perhaps most accomplished essay on the matter, "Folding in Time," Eisenman replaced Deleuze's "objectile" with the related and equally Deleuzian concept of "object-event": the moving and morphing images of the digital age break up the Cartesian and perspectival grids of the classical tradition, and invite architectural forms capable of continuous variation—forms that move in time.[14] Yet, in spite of the use of several formal stratagems, such as Thom's folding diagrams, the "folding" process remains purely generative,[15] and it does not relate to the visual form of any end product. Forms do not fold (actually, in all of Eisenman's projects featured in "Folding in Architecture," they fracture and break), because most buildings do not move. When built, architectural forms can at best only represent, symbolize, or somehow evoke the continuity of change or motion.

This argument would give rise to lengthy discussions in the years that followed,[16] but Eisenman's original stance was unequivocal: folding is a process, not a product; it does not necessarily produce visible folds (although it would later on); it is about creat-

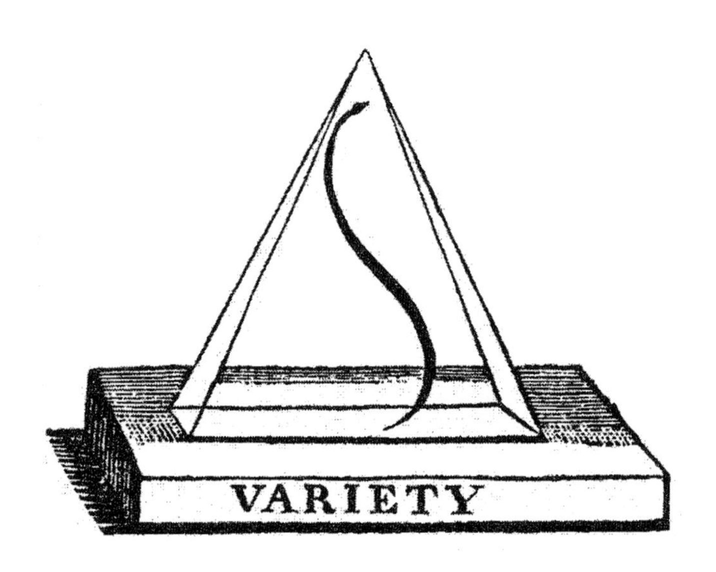

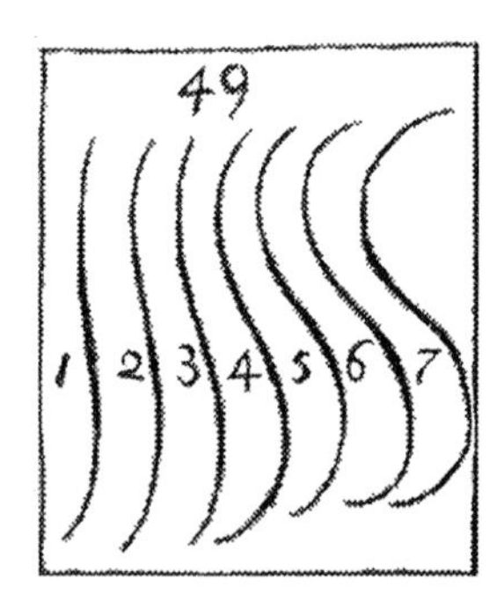

3.2 and 3.3
William Hogarth, *The Analysis of Beauty. Written with a view of fixing the fluctuating Ideas of Taste* (London: J. Reeves, 1753), title page, detail; and plate I, fig. 49.

ing built forms, necessarily motionless, which can nevertheless induce the perception of motion by suggesting the "continual variation" and "perpetual development" of a "form 'becoming'."[17] Again, art historians might relate such forms to a long tradition of expressionist design. Eisenman himself, at this early stage in the history of folding, defined folding as a "strategy for dislocating vision."[18]

In 1993, Lynn's preface to "Folding in Architecture" eloquently argued for continuities of all types: visual, programmatic, formal, technical, environmental, sociopolitical, and symbolic. The list of suitable means to this end is also remarkably diverse: topological geometry, morphology, morphogenesis, Thom's catastrophe theory, Deleuze's theory of the fold, and the "computer technology of both the defense and Hollywood film industries."[19] Nonetheless, even a cursory survey of the essays and projects featured in "Folding in Architecture" would puzzle today's readers. Most of the featured projects display a wide array of angles and fractures, ridges and creases, sharp-cornered pleats and lanky, bony, dissonant lines. The commentary blandly states that Gehry's irregular geometries were made possible by "3-D computer modelling,"[20] and digital technologies for design and manufacturing are mentioned by both Lynn and Jeffrey Kipnis as one tool among others that can help create "smooth transformations,"[21] but the one essay entirely devoted to computing, by Stephen Perrella, is on morphing and computer animation in the making of the movie *Terminator 2* (where the film's special effects director is quoted as saying: "we also used a program called Photoshop").[22] Lynn's commentary on Shoei Yoh's "topological" roof for the Odawara Municipal Sports Complex includes a stunningly perceptive and precocious analysis of the new tectonic, formal, and economic potentials brought about by the merging of computerized design, construction, and fabrication.[23] Yet, in spite of these isolated anticipations, "folds," circa 1993, were for the most part still intended and seen as angular pleats. With hindsight, digital technologies were then the main protagonist *in absentia*.

Meanwhile, however, the use of personal computers (and of the more powerful "workstations" of the time) was rapidly growing, and so were the graphic capabilities of the earlier mass-marketed software for image processing and computer-aided design. Most architects in the early nineties knew that computers could easily join dots with segments. But as CAD software quickly evolved, processing speed grew, and the price of new technologies declined, it soon appeared that computers could just as easily connect dots with continuous lines,[24] and sometimes extrapolate mathematical functions from them. Conversely, given a mathematical function, computers could visualize an almost infinite family of curves that share the same algorithm, of which the parameters can be changed at will. Architects discovered that smoothness, first defined as a visual category by the theorists of the picturesque at the end of the eighteenth century, was also a mathematical function derived from standard differential calculus.[25] Topological surfaces and topological deformations are equally described by mathematical functions—a bit unwieldy perhaps for manual operations, but already in the mid-nineties well within the grasp of any moderately priced desktop computer.

In this context, it stands to reason that Deleuze's theory and history of ontological continuity should inspire an architectural turn. Computers, mostly indifferent to queries on the nature of Being, can easily deliver tools for the manipulation of mathematical continuity. These new tools could be directly applied to the conception, representation, and production of objects. And they were. In the late nineties, Bernard Cache concluded that "mathematics has effectively become an object of manufacture,"[26] and Greg Lynn remarked that computer-aided design had "allowed architects to explore calculus-based forms for the first time."[27] To a large extent, our calculus is still Leibniz's, and the mathematical component of Deleuze's work on Leibniz, prominent but previously ignored, then sprang to the forefront—together with the realization that differential calculus was for the most part the mathematical language computers still used to visualize and manipulate all sorts of continuous forms. As Deleuze had remarked, Leibniz's mathematics of continuity introduced and expressed a new idea of the object: differential calculus does not describe objects, but their variations (and variations of variations). Deleuze even introduced a new term to characterize this two-tiered definition of the object—the "objectile," a function that contains an infinite number of objects.[28] Each different and individual object eventualizes the mathematical algorithm, or objectile, common to all; in Aristotelian terms, an objectile is one form in many events. Deleuze mentions Bernard Cache with regard to the mathematical definition of the objectile (adding that it corresponds to a new concept of the technical object, no longer mechanically made and mass-produced but digitally made and based on variations).[29] Cache's book *Earth Moves*, in which these arguments are developed, was published in English in 1995. The French manuscript, not published in French until 1997, was first drafted in 1983.[30]

So we see how a quest for formal continuity in architecture, born in part as a reaction against the deconstructivist cult of the fracture, crossed paths with the computer revolution of the mid-nineties and evolved into a theory of mathematical continuity. By a quirk of history, a philosophical text by Deleuze accompanied, fertilized, and catalyzed some stages of this process. Without this preexisting interest for continuity in architectural forms and processes—the causes of which must be found in cultural and societal desires—computers in the nineties would most likely not have inspired any new geometry of forms. Likewise, without

Yet, alongside this critique of technological change, which architectural invention may represent and even memorialize, actual technological change is happening (although not so fast as the exponential growth of digital technologies would have suggested in the nineties). Instead of focusing on one instance from a virtual series of many, the new technological paradigm is increasingly dealing with variations that can all be designed and fabricated sequentially: mathematical continuity in this case is set in a manufacturing series, not in a diachronic sequence, and used to mass produce the infinite variants of the same objectile—at the same unit cost as identical copies.

3.2 Standard

It is a tradition in all upscale or pretentious restaurants that the presentation of a bottle of wine should be accompanied by complicated rituals surrounding its tasting, climaxing with the consent, granted by the discerning patron, to the pouring of the wine that has passed muster. But the hieratic overtones of the ceremony should not disguise its prosaic, utilitarian origin. Wine is a delicate natural product: a work of nature and of vigilant craftsmanship. In spite of the amount of human intelligence and expertise brought to bear on the various phases of wine-making, each bottle of wine still counts, essentially, as an individual, unique, and unpredictable creation. As nature can be more or less merciful, and the maturing of wine may continue, unchecked, inside each corked bottle (cork is not a hermetic material, nor meant to be), trial before use is always advisable. When the wine tastes good, it may be served. No one, however, would perform a similar test before serving a bottle of Coca-Cola.

Coca-Cola is an industrial product. Its ingredients are set in a chemical formula; its production is entirely machine controlled and quality is tested throughout the process. As a result, each

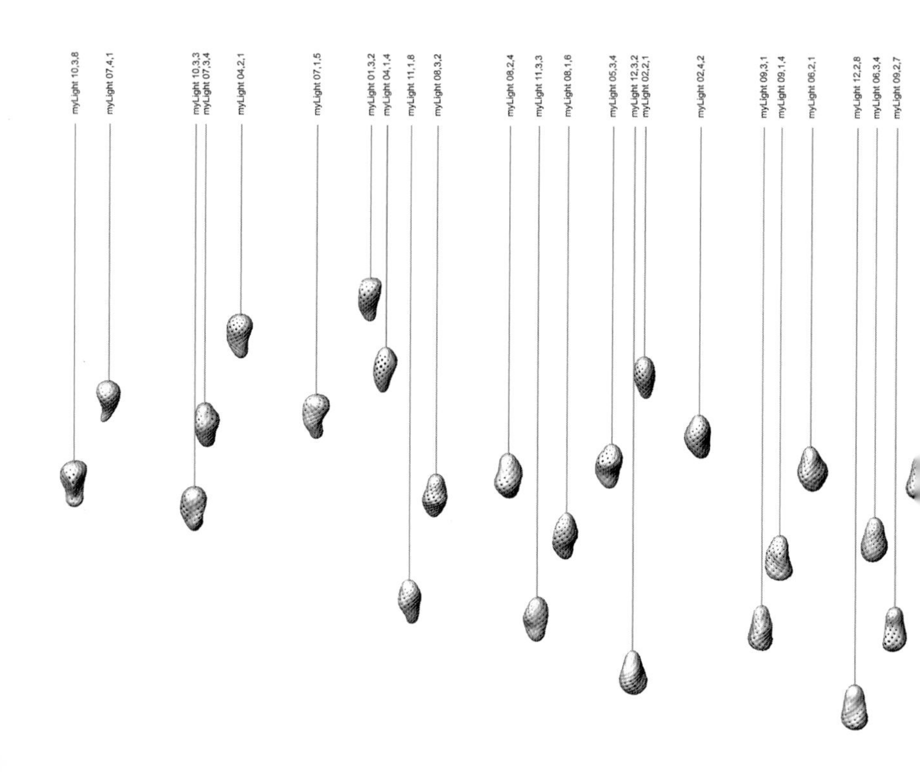

3.4
NOX/Lars Spuybroek, "myLight," 24 different nylon lamps for MGX Materialise, 2007.

bottle or can is expected to contain exactly the same product. As is well known, the Coca-Cola Company has occasionally tinkered with its magic formula, and it appears that the recipe may be marginally adjusted to take into account some regional preferences. Essentially, however, Coca-Cola is a standard product and it is expected to have a standard taste. When it does not, we know that the product is not the real thing. Unlike wine, Coca-Cola admits no nuances and no debate: either a glass of Coke is Coke, or it is not. If Heidegger had ever considered that something like Coca-Cola might exist, the quintessential *Gegenständigkeit* of this American drink would have supplied him with an endless source of inspiration for his musings on the debasement of the modern technical object (or, by contrast, on the "thingly" worth of Bavarian beer).

The transition from the organic variability of natural growth and artisanal manipulation to the mechanical standardization of modern mass production is particularly telling in the case of food and drink. Sigfried Giedion famously traced the beginnings of the modern assembly line to the standardization of manual biscuit-making in England during the Napoleonic Wars, and he has shown how, in this and other similar instances, the sequencing and standardization of the artisanal gesture preceded the introduction of machinery.[32] The formal arrangement of the "five rules for making good tea" and similar guidelines occasionally still printed on tea packages represent a similar transitional phase between traditional artisanal drinks and standardized, industrially produced ones. The corporate protocol for making Italian espresso followed by thousands of Starbucks "baristas" around the world (together with the espresso machines they use, developed for Starbucks by the Swiss company Thermoplan) may well represent the maximum degree of standardization compatible with the need for espresso coffee to be freshly made on site and on demand. But it is the history of whiskey in the twentieth century that offers one of the best examples of the modern rise of "the Taylorized beauty of the mechanical."[33]

Due to distillation, the chemical instability of spirits is less pronounced than that of wines; but, due to Prohibition, early in the twentieth century spirits in various parts of the world were bound to be—when found on the black market—artisanal, nameless, and very different in taste and quality. After the end of Prohibition, whiskey's rise to respectability in the United States required not only a recognizable brand name to stand as guarantor for the sanitary safety of the product, but also a recognizable, consistent taste for each brand. As every barrel of whiskey differs in its natural qualities from all others, product standardization was in this instance achieved through blending—an almost mathematical process whereby the statistical mean of many samples produces a relatively constant result.[34] The company that most deliberately pursued and profited from this strategy was, famously, Seagram—witness its commercial tag line "Say Seagram's and Be Sure" (first filed and registered in 1934–1935, and used through the 1970s). It appears that Samuel Bronfman himself (who bought the Seagram Company in the late 1920s) masterminded the transformation of a randomly changeable natural product into a safely predictable one, made to measure for modern marketing, modern taste, and modern consumption.[35] Architectural historians will not fail to notice an inner logic in that Seagram, a company whose commercial fortunes derived from intuiting and interpreting the spirit of modern standardization, should eventually build the most splendid monument to the idea of modern standardization in architecture, as it still stands on Park Avenue in New York.

In spirits, as in every other aspect of life and art, postmodern fragmentation has already undermined some of the staples of twentieth-century industrial modernism. Today's connoisseurs cherish the unique distinctiveness of single-barrel unblended whiskies and their quirky differences in taste, which not long ago would have put off high-modernist consumers. And, once again, digital technologies are poised to cater to this postmodern quest for individual variations—but through digitally controlled mass customization, not through manual craftsmanship.

As is well known, any color printer can imitate all the colors in nature by spraying combinations of a few primary colors on a white sheet of paper. Following the same principle, it would be theoretically possible to reproduce all known mineral waters in the world, for example, by mixing distilled water (the equivalent of a white sheet of paper) and the right amount of minerals, as determined by the chemical analysis of each natural spring. In the worst-case scenario, the cartridge of minerals inserted in this fictional mineral water fabricating machine should not contain more than the 117 or so chemical elements from Mendeleev's table, although for the purpose of making drinking water, fewer elements might be enough. And once the mineral cartridge and the distilled water holder have been loaded into the machine, production need not be limited to imitating existing waters—as endless new ones could be designed. As in the case of color printing, digital technologies would allow this machine to change the composition of the product anytime and at no additional cost, hence customizing each glass or bottle of mineral water in the same way a digital printer can sequentially print different colors, or different pictures, at will.

A project for the nonstandard mass production of one-of-a-kind, made-on-demand mineral water would be met with more or less respectful skepticism in a doctoral seminar or at a scholarly conference. Academia is often reluctant to catch up with reality. At the time of this writing, a similar machine is being tested by the very same Coca-Cola Company that, in the twentieth century, so iconically embodied the principles of mass production and standardization. According to current information, Coca-Cola's customizable fountain dispenser will operate from a touch-screen menu of preset options, but as the number of options is said to be in the hundreds, the dividing line between traditional multiple choice and seamless mass customization appears in this instance to already be on the wane.[36]

Constructions in steel or reinforced concrete tend to be less malleable and less prone to unlimited variations than music, pictures, or beverages; and all attempts at the mass customization of physical objects of design and manufacturing (or machinofacturing) are bound to be more limited in scope and import, and theoretically questionable. Yet the technical and cultural logics already at work are the same, and they will inevitably bring about some of the same consequences. The vertical integration of digital design and digital manufacturing, and the technical continuity between digital tools for visualization, notation, and fabrication, imply the elimination of most mechanical matrixes from the production process. That will spell the end of many basic principles of industrial economics. In the mechanical world, once a matrix is made, its cost must be amortized by using it as many times as possible. The economies of scale resulting from mass production are proportional to the number of identical copies that are obtained from the same mold: in mathematical terms, if the number of identical copies is infinite, the unit cost of the matrix is zero. The more you print, the less you pay per copy. Digital printing, however, does not work that way.

A laser beam must individually hit and mark all pixels on a printing surface every time anew, regardless of the printout that preceded it and of the one to follow. A metal plate that is used only once, to print one copy, will make for a very expensive print;

but a laser printer can print one hundred identical copies of the same page, or one hundred different pages, at the same cost per page. The same principle applies to all kinds of mechanical imprints, as well as to all processes where the identical repetition of the same sequence of actions, whether manual or mechanical, used to generate economies of scale. Under certain conditions, digital technologies can now deliver serial variations at no extra cost, and generate economies of scale while mass-producing series in which all items are different—but different within limits. On the eve of the digital revolution, Deleuze had famously anticipated this technological shift in his studies of difference and repetition, and, as mentioned above, his dual notion of "objet" and "objectile"[37] still aptly defines the basic principles of a nonstandard series.

A nonstandard series is defined not by its relation to the visual form of any constituent item, but by the variances, or differentials, between all sequential items in the series. A nonstandard series is a set in which each item has something in common with all others. In technical terms, all objects in a nonstandard series share some algorithms, as well as the machines that were used to process those algorithms and to produce the objects themselves. In visual terms, a nonstandard series comprises a theoretically unlimited number of objects that can all be different but must also all be similar, as the digital tools that were used to make them leave a detectable trace in all end products.

Algorithms, software, hardware, and digital manufacturing tools are the *new standards* that determine not only the general aspect of all objects in a nonstandard series, but also the aspects of each individual product, which may change randomly or by design.[38] Unlike a mechanical imprint, which physically stamps the same form onto objects, an algorithmic imprint lets outward and visible forms change and morph from one object to the next.

All items in a nonstandard series hence share the same style (in the original sense of the term, which referred to a writer's *stilus*, his tool—not his intentions). Today's *stilus* is not the style of the designer but the inevitable trace left by the software being used—which, just as Cicero's *stilus* was, is itself man-made and produced by specialized technicians.

A comparison between two instances of nonstandard series designed by two of the most alert interpreters of the new digital environment illustrates this point. Greg Lynn's series of ninety-nine teapots for the Italian luxury manufacturer Alessi,[39] and Bernard Cache's open-ended series *Projective Tables*,[40] feature individual items that are visibly different from, yet strikingly similar to, all the other items in the series. The peculiar *stilus* of each series derives in part from different technological platforms, but mostly from different adaptations of existing design software that were implemented by the two authors, each creating his own range of variabilities and his own set of self-imposed limits. Lynn favors software based on differential calculus; Cache developed an interface based on projective geometry. Both choices are arbitrary: each can be justified but neither is inevitable. In both cases the figural result is unequivocally distinctive: Lynn's differential calculus begets smoothness and continuous surfaces; Cache's projective geometry generates angular intersections of planes in three dimensions.

But the shift from mechanical to algorithmic reproduction also prefigures a parallel and equally crucial shift in our visual environment at large. We are leaving behind a universe of forms determined by exactly repeatable, visible imprints and moving toward a new visual environment dominated by exactly transmissible but invisible algorithms. To some extent, the recent discussions on the role of indexicality in contemporary architecture may in turn be a sign of the impending demise of identicality, on which much of our architectural and visual environments depend.[41] Most mechanically reproduced objects and forms are unmediated indices of the imprint that made them; most handmade works of the premechanical age, as well as most algorithmically generated items of the digital age, are not.[42] The erratic drift of manual copies may distort or confuse the sign of the original archetype and, as a result, conceal the identity of its author, or make it irrelevant; the unlimited variances of high-tech, digitally controlled differential reproduction may have similar consequences.

In the universe of mechanical *Reproduzierbarkeit*, the identification and, in turn, the meaning of forms depended on their identicality. In the new world of algorithmic, or differential, reproducibility, visual sameness is replaced by similarity.[43] Similarity and resemblance, however, are not scientific notions, and are notoriously difficult to assess and measure. The classical tradition, which was based on imitation, tried to nail down a workable notion of similarity almost from the beginnings of Western thought. Computer engineers and cognitive scientists today are trying to do the same—for the time being with less acumen than their classical and humanist predecessors. Indeed, there is an understandable, albeit ironic, vindication in the fact that some of our contemporaries, baffled by the apparently infinite range of variations generated by algorithmic reproduction, now attribute to mass-produced, identical copies some of the same nostalgic and "auratic" value that Walter Benjamin famously conferred upon premechanical, handmade originals early in the last century, when art theory first came to terms with the logic of identical copies.

The theory of nonstandard seriality was already fully inscribed in Deleuze's and Cache's seminal definition of the objectile, but it gained public and critical recognition only in the first years of the twenty-first century.[44] While some disagreements still persist on the very definition of "nonstandard" in architecture,[45] purely ideological concerns have recently fed an occasionally acrimonious debate on the matter. Even an unlimited supply of costless variations is hard to justify in the absence of a corresponding demand, and a technological revolution heralded by a cohort of ninety-nine teapots is easily disparaged. After all, making good tea is a fairly Taylorized and repetitive manual operation, and if the function does not change, there may be no reason why the form of the teapot should—nor any reason for its form to change algorithmically, incrementally, or differentially.

One could retort that experiments must often be performed and prototypes created at less than full scale, and that more crucial objects of manufacture than a teapot may follow (and indeed, some already have, although in a less glamorous fashion).[46] In many cases, nonstandard technologies and mass customization promise better, cheaper, and more suitable products for many. When the form of objects (from teaspoons to houses to cities) must follow human functions, standardized articles of mass production must aim at the average user or customer, and neglect the statistical margins, or "tails" in the bell curve of frequency distributions. When one size must fit all, those users or customers that don't fit—physically or ideologically, by choice or by necessity—must pay more to have what they need made to measure. Nonstandard technologies promise to alleviate this tax on diversity.

But this is the very reason why some may resent or reject nonstandard technologies. An egalitarian society posits some degree of equality in the forms and functions of all items of consumption and use, and mass customization goes counter to this ideological tenet. Technologies for nonstandard production may also appear to expand and multiply the offer of some commodities beyond necessity, hence fostering artificial demand and consumption.

The notion of mass customization was born as a marketing strategy well before the rise of CAD-CAM technologies,[47] and the offer of a range of options for the same product is an old gimmick of industrial mass production. Yet the range, scope, and theoretical import of seamless nonstandard seriality in a digitized environment should not be confused with the apparent multiplication of choice in a mechanical environment (often limited to marginal or cosmetic aspects of a product, such as color). The end of modernism, the end of industrial standardization, and the rise of neoconservatism occurred almost simultaneously with the rise of digital technologies and digital culture in the West; though it is difficult to establish a direct causal relationship between these events, it is easy to see why many that have taken a stance against the digital turn now see all new information technologies as agents of evil—or of financial capitalism. Using multiple forms for the same function may be vain, antisocial, and wasteful. Using the same form for multiple functions, on the contrary, is often ingenious, frugal, egalitarian, and virtuous. But in some cases it may also not work.

In the spring of 2005, some of Cache's *Projective Tables* were exhibited in an art gallery in central Paris.[48] The exhibit included a computer station where the customer was expected to set some variables in order to design the table of his or her choice; the resulting file would be e-mailed to the customer for further verification at home (for example, to double-check the measurements) before sending the order to the factory, where the parts would be laser-cut from a single wood plank and the table made to measure, and then shipped, either as an assembly kit or ready for use. While several incurious Parisians tried to buy the tables on display at the gallery and walk away with them, rather than design their own, the factory was not fictional, nor was the technology. When the whole system is put to use as planned, there is no reason why Cache's mass-customized tables should be more expensive than IKEA's mass-produced ones. Unlike IKEA's furniture, each of Cache's pieces is one of a kind and made to measure. Yet, just like IKEAs "Billy" bookcase, a standard in millions of homes around the world, Cache's pieces of furniture, while all different from each other, are easily recognizable once we have seen one.

And at the bigger scale of structural design, the tectonic implications of nonstandard technologies are equally vast and easier to measure. When form must follow structural constraints, such as load and stress, standardization inevitably begets oversizing and wasted material. This result is perfectly rational within the economy of a mechanical environment: in any given beam or cantilever, for example, there is generally only one section that is subject to the greatest stress, hence requiring the maximum size; all other sections could be decrementally smaller. However, beams with uniform sections are cheaper to manufacture than beams with variable sections, especially when they are mass-produced off site. Consequently, beams are often uniformly sized according to the single point of maximum load, because the savings generated by mass-producing oversized but standardized beams are greater than the cost of the excess material. When this pattern is repeated in all parts of a large structure, the inevitable result is dumb structural design and a waste of building material. Indeed, both results characterized standard civil engineering throughout most of the second half of the twentieth century. But this trend can now be reversed. Thanks to digital technologies, it is possible to envisage a new generation of nonstandard load-bearing components, both mass-produced and made to measure, and using no more material than is required at any given point of a structure.

Digital tools could also enable the structural use of new, nonlinear materials—or, to be precise, materials characterized by nonlinear elasticity, such as plastics or ceramics—and a nimbler exploitation of organic building materials like timber or stone, which may be structurally unpredictable due to natural variations.[49] Nonstandard technologies could interact with such irregularities, and adapt form and design to the variability of nature almost as aptly as artisanal manipulation once did. Using fine-grained load-bearing calculations, three-dimensional geometrical notations, and robotic fabrication, digital technologies could also enhance the structural and formal potentials of traditional building technologies, such as masonry.[50] If some of these promises were to come true, structural design could once again become an art, as it often was in the past, when building materials were rare and human intelligence abounded. For most of the last century, the reverse has been the case. The preindustrial world always knew that all raw materials, including building materials, were hard to come by. Recent global events have reminded the industrial world that some basic commodities are in limited supply. Today, ingenuity in structural design—which always implied some form of elegant parsimony in the exploitation of natural and human resources—is more than a matter of virtuosity or pride for designers and makers; it is also a matter of social responsibility for all users.

The rise of the nonstandard paradigm implies the reversal of some technical, economic, social, and visual principles that have characterized the mechanical age for most of the last five centuries. But the transition from identical to differential reproduction also revives some aspects of a pre-Albertian technocultural environment, a world of shared technical lore and collaborative making, while at the same time evoking a neo-Scholastic frame of mind, based on algorithmically defined, fixed genera and endlessly drifting visual species.[51] This is a world that one of the greatest moralists of the twentieth century, Lewis Mumford, prophesied as forthcoming—anticipating, in the wake of the Great Depression, the redemption of a new technological future that would stamp out the paleotechnical, mechanical age he decried as the source of all evil.[52] This is also a world that the greatest moralist of the nineteenth century, John Ruskin, situated in the past, in the golden age of Gothic artisanship, when all work was man-made and still inspired by the Lamp of Life, before the slavish art of the Renaissance turned artisans into machines, and European architecture lost all light and all life, as Ruskin thought, forever.[53] But for these same reasons the emerging nonstandard environment is bound to be meaningless, or worse, for many of today's architects and critics who trained in the credos of twentieth-century modernism—or are nostalgic for it. To take only one of the best-known examples: Rem Koolhaas's theory of the "generic" champions the exact opposite of the nonstandard specificity of form; his theory of "bigness" is conspicuously contrary to nonstandard adaptiveness, and the famed four Koolhaasian standards, "S/M/L/XL,"[54] unapologetically celebrate just what they stand for: the standards of an age gone by.

3.3 Agency

Sometimes the West is obsessed with the idea of progress, sometimes with the idea of decline, and sometimes both notions fall off the radar entirely. The bakery that served my neighborhood in the small Italian town where I grew up was called, I remember vividly, Il Forno Moderno (The Modern Bakery). Italy was highly confident of its future in the late fifties and sixties; that general optimism evidently included the expectation that modern technology would deliver more and better bread. Later the mood changed. As far as I can remember, the bakery went out of business in the early eighties, and after a while someone reopened it under a new name—L'Antico Forno (The Old Bakery). When I last visited the place the bakery had disappeared, and the site was occupied by a cell phone store. It wasn't called either modern or old; the apparent name of the business, or at least the only visible name in the shop window, was "Global Roaming" (in English).

The digital revolution that marked architecture at the end of the twentieth century may have been the first self-proclaimed revolution in recent Western history to take place for the most part without, and outside of, any established philosophy of history (to revert to my bakery analogy: no future, no past—just "global roaming"). This may seem a contradiction in terms, as the notion of a revolution implies that something is being disrupted; and the primary object of a revolutionary disruption, in the Hegelian tradition, used to be an historical process. But a revolution is a modern ideological construct. If you try to make one in a postmodern environment, odd things may happen—as they did in this case.

Postmodernism in architecture is often associated with the revival of iconicity, historicism, and symbolism. This, however, was only one brand of postmodernism, the one Charles Jencks invented specifically for architectural consumption, almost simultaneously with what we should perhaps call—to differentiate it from the former—the philosophers' postmodernism, which had the same name, and similar causes, but different effects. In Jean-François Lyotard's version, postmodernism proclaimed the "fragmentation of all master narratives," first and foremost the most pervasive of recent Western "strong referentials," which was Hegel's philosophy of history. The ensuing corollary, the end of history, may have been first announced by Jean Baudrillard a couple of years later (together with the end of mimesis, representation, art, society, and more, including the end of the

3.5 and 3.6
Gramazio & Kohler in cooperation with Bearth & Deplazes, facade of Vineyard Gantenbein, Flaesch, Switzerland (2006): detail of the wall; the bricklaying machine. Copyright Gramazio & Kohler, Architecture and Digital Fabrication, ETH Zurich.

world).[55] Baudrillard would eventually partly change his mind, at least on the matter of history,[56] but by then the beast was out of the cage: the expression was famously appropriated by the neo-conservative ideologue Francis Fukuyama and, in the aftermath of the fall of the Berlin Wall, the two versions of the end of history (the postmodernist and the neoconservative) merged into a general perception that directional, teleological history (the Hegelian view of a rising line of progress moving from A to B) had run its full course. For the NeoCons, history had reached its final destination, and from there could go no further; for the PoMos, the very notion of a single line of progress had become meaningless and irrelevant.[57]

After the fall of socialism in Western Europe, the "dividends of peace," as they were called at the time,[58] kept piling up during the relatively peaceful and prosperous *fin de siècle* that lasted from November 9, 1989,[59] to September 11, 2001, as the NASDAQ and "irrational exuberance"[60] surged, and the digital revolution in architecture "folded away from deconstructivism,"[61] as some of us may remember. This was precisely the time when the Deleuze-Eisenman connection crossed paths with the new tools for computer-aided design, new theories on folds, blobs, and topological geometries flourished, and the general awareness of a digital revolution in architecture started to take root.[62] Yet, in spite of the general excitement for the extraordinary advances in all branches of digital technologies, this may well have been the first wave of technological progress in recent history to unfold in the absence of any related ideology of progress, hence the first self-defined revolution to occur in an historicist vacuum.

This context may explain some of the peculiarities of the digital revolution of the nineties. Any revolution, even a technical one, by definition changes the course of history, but this one had no clearly identifiable, preexisting course of history to refer to or call into question. Consequently, in true postmodern fashion, it had no preset destination: no target, as it were, and almost no end in mind. A revolution without an enemy is a solution without a problem, and even in a postmodern environment, a revolution without a vision for the future is an anomaly. But the posthistorical environment where the first digital revolution in architecture emerged and thrived has had more crucial and long-lasting consequences.

As Fukuyama had foretold, the collapse of many modernist "strong referentials" would bring about a decline in ideal tension, ambition, and drive, paving the way for an "old age of mankind" marked by a rarefaction of collective and personal expectations; in this gray dull universe of "men without chest" (a notorious expression Fukuyama borrowed from Nietzsche), a few surviving "men in full"[63] would inevitably look for other ways to indulge their will to power. In the absence of war, they might find suitable outlets in corporate capitalism and in the financial markets, where they could play "masters of the universe," but with fewer risks than medieval warlords took on. From Vitruvius's Dinocrates to Ayn Rand's Howard Roark (and beyond), expanded egos and megalomania have always been an inevitable component of architecture, but digital *Übermenschlichkeit* in the nineties went on to acquire some distinctive traits. The new nonstandard environment was often seen as a liberation from the ideological straitjackets of modernism—including social and collective responsibility. As many argued at the time, the new digital tools would finally liberate creative forces that technology and society had long constrained, enabling the expression of nonstandard individualities, differences and variations that older technologies could not support, and older societies would not tolerate. This myth of a new Prometheus digitally unbound, asserting his right to digital *Bildung* against all comers, was a strong component of the digital movement in the nineties (and fragments of it survive to this day, particularly in the work of some architects and designers, now famous, whose digital style was forged in that context).

"I do it because I can": as digital tools are fantastic amplifiers of the power of making form, digital technologies were crucial players in the technocultural formation of this earlier postmodern, posthistorical environment. Nor was this quintessential empathy between digital technologies and postmodernity limited to architecture. In the nineties it was capitalism itself, in a sense, that went Deleuzian.[64] Lyotard's "fragmentation of master narratives" turned into a fragmentation of the marketplace; Deleuze and Guattari's "thousand plateaus" became a thousand market niches; these, in turn, created a demand for mass customization—which, as it happens, is what digital technologies can best provide. This new pattern of "made-on-demand," atomized production and consumption is today widely seen as an as an emerging and possibly disruptive economic model.[65] At the same time, the new information technologies have favored the decoupling between the "real" economy and the financial markets, and hastened the virtualization of financial transactions.[66] In all such instances, digital technologies and various cultural aspects of the first age of postmodernity interacted with and fed back on one another. This interplay is even more conspicuous in the case of architectural design, as the formal revolution of the nineties resulted from a match between the digital supply of a new generation of variable forms and the postmodern demand for individualized variability in form-making.

As we now know, the decade of the end of history would last twelve years, and would end in disorder and disaster. In the new, contrite mood that prevailed after 2001 some digital pioneers repented and some outright recanted.[67] Other protagonists of the first digital wave kept doing their jobs—sometimes with more experience and less theoretical boldness, or none, and this perceived maturity has recently garnered critical recognition.[68] But not all digital technologies shut down or froze in the wake of the dot-com bust of 2000–2001. In fact, technology kept changing, and today, new technosocial developments invite new uses and challenge new users.

CAD-CAM of the nineties was mostly based on controlled, proprietary networked environments. The emphasis was then on the vertical integration of digital design and fabrication, and on the potential of this new technology for the production of nonstandard, serial variations. But in recent years the networked environment has evolved from earlier, mostly monodirectional information technologies ("one-to-many") to a fully symmetrical, bidirectional informational framework. This technical development is being exploited for a variety of purposes, some purely technological (such as P2P and distributed processing networks, or cloud computing), some with vast social implications—and indeed, some of this software is also called "collaborative" or even "social" software. As every node on the net can now be both a receiver and an emitter of information, many users have started to use the web to do just that. The notion of "user-generated content" implies that in many cases the user may still be discernible from official producers of content, but one can think of many other cases where the present web (also known as the Web 2.0)[69] is fast becoming a flat, isotropic platform where the traditional divide between sender and receiver, or author and audience, is fading. In line with this general trend, the emphasis in architectural design has shifted from earlier, tectonically oriented applications to the interactive, "horizontal" integration of the design process itself—and of the various categories of actors, agents, and agencies it may include.

As of this writing, agency, interactivity and participation are the catchwords of the day, and the architectural discourse on these matters is now as intense as the discourse on hypersurfaces, nonstandard, and topological geometries was ten years ago.[70] Interactive connectiveness may include human participants as well as machines of all sorts, and until recently technical interaction between networked machines[71] attracted more interest than digitally enhanced social collaboration between humans. With some reason: responsive technical environments may include exciting new architectural features and gadgets, whereas teamwork in architecture is hardly a hot new topic. Digitally enhanced or not, architectural design has always been a delicate act of negotiation and balance between many participants, personalities, and committees; between the individual and the collective. The pendulum may swing both ways: in the not so distant past, public participation in design was actively pursued by many socialist architects in Europe, and in America by community activists and advocacy planners.

But the new forms of digitally supported social participation in decision-making are significantly different in spirit from traditional, consensus-seeking modes of "design by committee." Interactive digital versioning (as supported by wiki technologies, for example) posits a never-ending accrual of independent, diverse, and individual edits and changes, where consensus is neither sought after nor ever achievable.[72] Evidently, architectural end products must at some point be built in solid, nonvariable, and permanent form—a limitation that does not apply to media objects that can keep changing *ad libitum*. But this exception is less determinant than it would first appear, as architectural notations can morph and change even when reinforced concrete, once cast in a mold, cannot. The whole theory of digital differentiality in architecture is predicated upon notational (i.e., informational) variations—on objectiles, not on physical objects. And, indeed, the "social" maelstrom of the Web 2.0 has by now reached even mainstream architectural and engineering software, with some already notable consequences.

Recent developments in building information modeling (BIM) emphasize the participatory potential of a technology that was originally devised to manage sets of databases during the design and construction of building components.[73] Conceivably, three-dimensional BIM platforms may soon be used to create and maintain digital models of even complex buildings throughout their life cycle—from conception, design, and construction to operation, maintenance, and disposal. For centuries traditional scaled models have provided a venue for information exchange and discussion on various aspects of buildings, or any another technical object under design or development; realistic 3D models in particular have often been used to consult nontechnical agencies—typically, patrons, customers, councils, committees, or even the general public.[74] Given their unequaled potential for visualization, immersive simulation, and interactive feedback, digital technologies could greatly enrich these traditional tools for publicity and participation in design.

The main obstacle to these developments is not technical. User-generated content has already been both praised and blamed for the perceived obsolescence of some old media. In the context of the ongoing discussions about the so-called Web 2.0, and given the sometimes messianic, sometimes apocalyptic premonitions of a digitally induced disappearance of traditional authorship, architects have additional reasons to fear developments that may amplify the role of other participants in the design process—to their detriment. These preoccupations, however, may be overblown when seen in historical perspective.[75] For what is at stake today, and what may indeed be lost, is not

The Author, as a timeless category of the spirit, but a very technologically specific kind of author. The falling star is, simply, the author of reproducible identical copies—a vast and influential category for sure, but not an indispensable one.

Take, for example, music, a medium that has been deeply affected by the new digital environment. One may regret the loss of the almost inviolable integrity of LPs and CDs, which were common before the age of streaming or downloadable music. Records and discs were authorial recordings of sound, intended to exist always and only in that one authorized version. Mozart and Beethoven, however (or even Arnold Schoenberg or Luciano Berio, for that matter), never published sounds: they wrote notations, which were recorded and transmitted in writing, and became music only when performed. And every performance, or enactment, of the same musical score was bound to be different (within limits) even when the composer was playing or conducting in person. The notion of music as an authorial, autographic, identical reproduction of sound is only a recent historical acquisition of the mechanical age. For centuries, even when it was formally written, music itself was a permanently variable medium, endlessly drifting and morphing; interpreted, edited, reworked, and remixed by countless performers, composers, and amateurs.[76] There is, of course, a difference between changing interpretations of the same musical notation (traditional variability) and the random mixing and matching of snippets of digitally recorded sounds (digital differentiality). Yet in the case of music, too, today's transition from mechanical to digital media once again implies the transition from a standardized world of ready-made items for all (from the same printed LP) to a new variable environment of customizable items for each (from each personal iPod).

Modern architectural authorship came into being only with the rise of what I have called the Albertian paradigm—the definition of architecture as an allographic art, and of building as the notationally identical copy of a single, authorial act of design. Even though it was never really fully implemented, not even in the twentieth century, this paradigm has nevertheless inspired most of Western architecture for the last five centuries, and it is at the basis of the dominant legal framework that still regulates the global practice of the architectural profession. It was at the dawn of modernity that Alberti forcefully shaped that pervasive and essential tenet of Western humanism, asserting that works of the intellect, including architectural works, have one author, and one archetype, which executors are required to reproduce identically and prohibited from altering. And this is the paradigm that recent developments in digital technologies are now phasing out.

Digital technologies are fast defrosting several classes of media objects that mechanical technologies froze over the course of the last five centuries—from woodcut prints and Gutenberg's press to more recent electromechanical or chemical-mechanical technologies for the recording and transmission of sounds and images. Digitally supported texts, music, and pictures can now start to drift again, as they did before the brief interlude of the mechanical age. In architecture, the vertical integration of computer-based design and manufacturing is giving rise to new forms of digital artisanship, narrowing the Albertian divide between conceivers and makers. Likewise, the digitally enhanced horizontal integration of actors and agencies in the design and production process is already challenging the modern notion of the architect's full authorial control and intellectual ownership of the end product. New digital platforms for open-ended, interactive collaboration may beget endless design variations, revisions

The open source movement, which predates current Web 2.0 developments, famously pioneered the same principles of commonality and the spirit of collective creative endeavor now pervasive in the world of social media and user-generated content. Not coincidentally, reference to Gothic building is prominent in one of the foundational texts of the movement.[79] Nor has the analogy between digitally supported collaboration and preindustrial craftsmanship escaped more recent, ideologically motivated scrutiny.[80] Indeed, a broad and vague reference to Gothic art and to the Middle Ages has inspired various aspects of contemporary digital culture almost from the start, including in its more popular aspects, and this trend has long been acknowledged by media scholars.

In the case of architecture, however, deeper and more specific analogies raise more momentous implications.[81] Contemporary Heideggerians like to oppose the communal, almost mystical binding and gathering value of all actively man-made Things (*Ding*) to the external, muted neutrality of technical objects of manufacturing (*Gegenstand*). The often not-so-hidden agenda of some of today's belated phenomenologists does contemplate the suppression of most existing technologies, yet the ongoing mutation of our technological universe from the mechanical to the digital may fulfill, oddly, part of the same technophobic program. Digital tools no longer need to separate the thinker and the making; on the contrary, if pertinently put to use, digital technologies may reunite most of what mechanical technologies had alienated—including the various communities that in the premechanical world were associated with, and dignified, all Things handmade. Bruno Latour recently envisaged a similar shift toward a new Gathering of Things—the knotting, binding, and linking together of a new class of objects and subjects (what Latour calls "matters of concern").[82] There are different ways of

4

EPILOGUE: SPLIT AGENCY

Historians of the present should never stop writing. But all works destined for mechanical reproduction must be permanently arrested at some point in time, when authors confide them to the copying machine—literally, or figuratively. This book is no exception. The preceding chapters were written before (in some cases, shortly before) the financial crisis that came to a head in the fall of 2008, and is still unfolding at the time of this writing, fourteen months later. As already occurred in 2000–2001 (and many times before), once again a global financial and sociopolitical upheaval is prompting a reset of our perception of technological change.

For the last five years or so, the participatory use of digital technologies has been adopted with equal enthusiasm by visionary libertarians and liberal neoconservatives—a reminder of the dual and possibly ambiguous cultural allegiances of digital interactivity itself. Historically, the open source movement is a descendant of the rebellious and often anarcho-socialist hacker culture of the eighties and nineties.[1] And only a few months ago, one of the founding fathers of the Web 2.0 compared today's collaborative digiteracy to a "digital socialism" of sorts.[2] Yet the very principles of today's collaborative web evoke a starkly different ideological provenance. Wikipedia functions, when it does, because the summation of a very large number of individual contributions, endlessly revising one another, produces a collective body of knowledge that transcends the errors and limitations of each participant. But tapping the wisdom of crowds, whether digitally or otherwise, is hardly a new idea. The invisible but self-correcting hand of the Web 2.0 is next of kin to the "invisible hand of the market" that, according to Adam Smith's classical theory of liberalism, transforms the self-interest of each market participant into the superior wisdom and righteousness of the market itself. And when, thanks to digital technologies, the markets can include a theoretically unlimited number of participants and an equally infinite number of near-instantaneous, frictionless transactions, free markets become ideally infallible—or so some economists were inclined to think until recently.[3]

As the financial bubble and the architectural blob shared many scripts and codes, and were in fact both made possible by the same technologies, the burst of the former did nothing to improve the reputation of the latter. Of course, digital tools in architecture meant and mean more than blobs. But the very same communal spirit that pervaded the Web 2.0—in its multifarious, libertarian, as well as neoliberal manifestations—is now in doubt, and possibly retrenching.[4] This is happening at a time when many new digital tools inspired by the ideals of the Web 2.0 are going mainstream: distributed peer production, once limited to media objects (texts, images, music), is now being applied to the actual manufacturing of physical hardware;[5] this postindustrial, "do-it-yourself economy" of digital making has even inspired a recent work of science fiction—and a prize-winning entry by Greg Lynn for the 2008 Venice Biennale.[6] In architecture, various software packages for building information modeling are quickly becoming global and almost inevitable industry standards for architects, contractors, and even government agencies. And the potentials of these new tools go well beyond the mostly bureaucratic purposes to which they are presently confined.

Soon BIM applications will be able to create and maintain a permanent, interactive digital doppelgänger of each object of design. Using similar technologies, design will increasingly be achieved through visual and pictorial interfaces, such as mimetic (perspectival) renderings and even immersive environments, to the detriment of the traditional formats of architectural notations (plans, elevations, and sections).[7] These developments will offer new and hitherto unimaginable venues and possibilities for openness and participation by even nontechnical agents in all stages of design.

Architects may be understandably reluctant to surrender their hard-won authorial identity to a new digital public. Yet a similar devolution is already happening in the opposite direction, as often-unwitting architects are enticed into becoming clients (hence the subordinate public) of design agencies of a higher order. Given the complexity and cost of the new digital tools and the import of the tasks they can accomplish, big offices and corporations have recently started to offer full BIM services to smaller architectural firms. Independent architects may soon avail themselves of BIM design platforms on a pay-as-you-go basis, leaving the service provider in charge of the technical side of the process, up to and including the final project delivery.[8] And at the smaller scale of rapid prototyping, some high-tech manufacturers have for some time been marketing their integrated design and fabrication tools to a select public of digital designers, who lease the company's proprietary file-to-factory technology to develop and produce their own creations.[9] These *objets d'art*—which, not surprisingly, often look somewhat similar to one another—may be sold under the names of their respective designers, but the key coauthor of these works is in fact the company that owns the technology used to design and make them.

Architects would do well to heed the risks and paradoxes inherent in such new forms of split agency. All parametric design inevitably entails two levels of authorship: at one level, the primary author is the designer of the generic object (or objectile: the program or series or generative notation); at another level, a secondary author specifies (in the Aristotelian sense of the term) the generic object in order to design individual end products. In this dialectic, the secondary is to the primary author what the player in a video game is to the video game's designer: each gamer invents (or, in a sense, authors) her or his own story, but playing by the rules of the game and within an environment conceived by someone else. As Janet Murray remarked long ago, the player in a digital video game—an "interactor" rather than an author—exerts only a limited and ancillary form of agency.[10]

Architects that by choice or by necessity intervene in someone else's digital design environments are to some extent only secondary authors—end users and not designers.[11] As the digital turn is reshaping the Albertian and modern terms of architectural agency, and architects adjust to a new and untested authorial environment, many transitional hybrids between the old model and the new are likely to be tried out. But the spirit of the game should not be lost, regardless of the roles that architects may be called to play. Open-endedness, variability, interactivity, and participation are the technological quintessence of the digital age. They are here to stay. And soon designers will have to choose. They may design objects, and then be digital interactors. Or they may design objectiles, and then be digital authors. The latter choice is more arduous by far, but its rewards are greater.

Objects belong to the old, mechanical world of identicality and products, of centralization and authority. Objectiles belong to the new digital world of variability and process, of participation and community. The old world offers a multiplication of choices in an ever-growing catalog of ready-made products, hence consumerism. The new world promises seamless, on-demand customization through interactive decision-making, hence—ideally—social responsibility in design, as well as parsimony in the use of natural and human resources. To embrace digital authorship in full, however, designers will need to rise to the challenge of a new, digitally negotiated, partial indeterminacy in the process of making form. And this will not be easy, as no architect was ever trained to be a generic author—nor, most likely, ever had the ambition of becoming one.

As in all times of major upheavals, some fear of the future is warranted. But a new algorithmically driven, interactively generated visual environment need not be a gaudy dump of cyberpunkish, unreadable formlessness (as science fiction writers often imagine), nor a bland, design-by-committee, regression-to-the-mean, technocratic landscape of benign formal negligence. As in all tradeoffs, some things will be lost and some gained. Architects may have to relinquish some control on specific end products, but they will acquire full control of generic objects. As the programmers and masterminds of nonstandard series, they will preside over the full extent of general, genetic, or parametric visual environments at all scales, from teaspoons to cities (or, today, from nanotechnologies to global territories). An objectile's designer can script programs capable of prodding the intelligence of each individual end user, marshaling the talent of an unpredictable community of future interactors, curbing the unavoidable foolishness, and allowing for the removal of errors. A generic environment is not uncontrollable. Quite to the contrary, every open-ended, generic environment is strictly controlled by the very same generative rules that make it possible.

The modern process of architectural design, and the architect's authorial role in it, may not survive the digital turn. Yet, as architecture preexisted both the invention of the Albertian author and the rise of mechanical copies, neither may be indispensable to its future. The post-Albertian architecture of our digital future will have something in common with the pre-Albertian architecture of our artisanal past, but this does not mean that digital architecture might or should look Gothic—nor any other style. For one thing, premechanical classicism was as much parametric and generative as medieval stereotomy. But this is not the point. Similar processes do not necessarily beget similar shapes. Understanding these processes, on contrary, will help us shape better things.

NOTES

CHAPTER 1

1. Now partly available on YouTube (<www.youtube.com/watch?v=iRzr1QU6K1o>, accessed December 17, 2009).
2. Most developed countries had left the gold standard in the course of the 1930s, as a consequence of the Great Depression. The Bretton Woods Agreements reintroduced stable exchange rates between currencies and the U.S. dollar, and between the U.S. dollar and gold; the system collapsed when Nixon unilaterally suspended the convertibility of the dollar into gold at a conventional, fixed rate.
3. Jean Baudrillard, *Simulacres et simulation* (Paris: Galilée, 1981), 70.
4. Jean-François Lyotard, *La condition postmoderne* (Paris: éditions de Minuit, 1979), 31. Lyotard's original expression was "décomposition des grands Récits."
5. Gilles Deleuze and Félix Guattari, "November 20, 1923: Postulates of Linguistics," chapter 4 in *A Thousand Plateaus: Capitalism and Schizophrenia*, trans. Brian Massumi (Minneapolis: University of Minnesota Press, 1987), esp. 90–97. Originally published in French as *Capitalisme et schizophrénie 2, Milles plateaux* (Paris: éditions de Minuit, 1980). See Reinhold Martin, "Environment, c.1973," *Grey Room* 14 (Winter 2004): 93, n58. Deleuze and Guattari's reference to economic history (namely to the replacement of the Reichsmark on November 20, 1923) gives the title to the aforementioned chapter of *A Thousand Plateaus*, but it is anecdotal with regard to their general arguments.
6. First published in 1959.
7. E. Victor Morgan, *A History of Money* (Baltimore: Penguin, 1965), 28.
8. Sometimes an additional three- or four-digit card security code (CSC) is added to the main string of sixteen. Before today's "chip and PIN" credit or debit cards,

earlier generations of cards (still popular in the U.S.) required the customer's signature on a paper form for each transaction, as well as on the verso of the card for verification. For reasons unknown, but probably dependent on tradition, nostalgia, or technological inertia, most credit card issuers still require that chip and PIN cards be signed by their owner.

9. See Alice Rawsthorn's column on design in the *International Herald Tribune* (last to date "What defies defining, but exists everywhere?" August 18, 2008; also available online at <www.nytimes.com/2008/08/18/arts/18iht-DESIGN18.1.15327742.html>, web site accessed December 23, 2009); Umberto Eco, "On the End of Design," *Lotus* 138 (2009): 115–120.

10. The notion of "differentiation" is key to Lynn's theory. Lynn first used the term in its original mathematical sense (as in "differential variations," and "geometrical ... differential transformations and deformations"). See his "Multiplicitous and Inorganic Bodies," *Assemblage* 19 (December 1992): 32–49, esp. 35. Lynn then generalized the same notion, extrapolated from differential calculus, to describe formal continuity and smooth surfaces (see the "continuously differentiated surfaces" in Shoei Yoh's roof structures, in Greg Lynn, ed., "Folding in Architecture," special issue [AD Profile 102], *Architectural Design* 63, nos. 3–4 [1993]: 27; see also "Differential Gravities," *ANY* 5 [March-April 1994]: 20–23, repr. in Lynn, *Folds, Bodies and Blobs: Collected Essays* [Brussels: La lettre volée, 1998], 102; "composition through differentiation" and "continuous, indivisible differentiation," in "Body Matters," *Journal of Philosophy and the Visual Arts* 5 [1995], repr. in *Folds, Bodies and Blobs*, 140; and "differential complexity" and "complexity, continuity and differentiation," in "Blobs," *Journal of Philosophy and the Visual Arts* 6 [1995], repr. in *Folds, Bodies and Blobs*, 158, 168). The same mathematical principles of continuity were eventually related to tectonics and fabrication in terms that more specifically characterize nonstandard seriality: see for example Lynn's reference to "a differential approach to series and repetition" in *Animate Form* (New York: Princeton Architectural Press, 1999), 33; and Lynn's editorial introduction to the reprint of "Folding in Architecture" (London: Wiley-Academy, 2004), 12 ("from the identical asexual reproduction of simple machines to the differential sexual reproduction of intricate machines").

11. See John Battelle, *The Search: How Google and Its Rivals Rewrote the Rules of Business and Transformed Our Culture* (New York: Portfolio, 2005).

12. Stephanie Clifford, "Billboards That Look Back," *New York Times*, May 31, 2008. Howard Rheingold had already described a similar technology in 2002, when it was apparently being developed by IBM's Almaden Research Center. See Rheingold, *Smart Mobs: The Next Social Revolution* (Cambridge, MA: Perseus Publishing, 2002), 94, 187, 193.
13. Mario Carpo, "Pattern Recognition," in Kurt W. Forster, ed., *Metamorph: Catalogue of the 9th International Biennale d'Architettura, Venice 2004*, 3 vols. (Venice: Marsilio; New York: Rizzoli International, 2004), 3:44–58, esp. 45n1.
14. This applies only to meanings derived from visual sameness (as in the case of some indices) or from visual similarity (icons); it does not apply to "symbolic" meanings (as defined by Peirce's theory of signs).
15. Here and elsewhere in this text, "mechanical" refers specifically to the second of these three ages, whereas "machines" may apply both to mechanical and to digital devices.
16. See for example Lev Manovich, *The Language of New Media* (Cambridge, MA: MIT Press, 2001).
17. See for example William M. Ivins, Jr., *Prints and Visual Communication* (Cambridge, MA: Harvard University Press, 1953); or more recently, Friedrich A. Kittler, "Perspective and the Book," *Grey Room* 5 (2001): 38–53, first published in German as "Buch und Perspektive," in Joachim Knape, ed., *Perspektiven der Buchund Kommunikationskultur* (Tübingen: Hermann-Arndt Riethmüller, 2000), 19–31.
18. In Bruno Latour's terms, "immutable mobiles." See Latour, "Drawing Things Together," in Michael Lynch and Steve Woolgar, eds., *Representation in Scientific Practice* (Cambridge, MA: MIT Press, 1990), 19–68; esp. 34–35, 47; an earlier version was published as "Visualization and Cognition: Thinking with Eyes and Hands," *Knowledge and Society: Studies in the Sociology of Culture Past and Present* 6 (1986): 1–40. For further reference to the seminal works of William Ivins, Walter Ong, and Elizabeth Eisenstein on this matter, see Mario Carpo, *Architecture in the Age of Printing* (Cambridge, MA: MIT Press, 2001), revised and translated from *L'architettura dell'età della stampa* (Milan: Jaca Book, 1998). See also Christopher S. Wood, "Notation of Visual Information in the Earliest Archeological Scholarship," *Word & Image* 17, nos. 1–2 (2001): 94–118.
19. See esp. Carpo, *Architecture in the Age of Printing*.

20. In order to look identical to one another, some individual parts had to be made differently—witness the theory of optical corrections in antiquity and in the Renaissance. Moreover, in the classical tradition identical copies often implied proportional, not mensural, identicality (see below, section 1.5). The classical tradition dealt with visual sameness, not with standardized mass production.

21. Nelson Goodman, *Languages of Art: An Approach to a Theory of Symbols*, 2nd edn. (Indianapolis: Bobbs-Merrill, 1976), 122, 218–221.

22. Spiro Kostof, "The Practice of Architecture in the Ancient World: Egypt and Greece," in Kostof, ed., *The Architect: Chapters in the History of the Profession* (New York: Oxford University Press, 1977), 3–27, esp. 7–8 on the Turin papyrus.

23. See Jens Andreas Bundgaard, *Mnesicles: A Greek Architect at Work* (Copenhagen: Gyldendal, 1957); Kostof, "The Practice of Architecture," 12–16, 26–27; J. J. Coulton, *Ancient Greek Architects at Work: Problems of Structure and Design* (Ithaca: Cornell University Press, 1977), 51–73, esp. 53–58 on Greek *syngraphai* (specifications or technical descriptions), *paradeigmata* (full-size three-dimensional examples), and *anagrapheis* (textual descriptions or planar full-scale templates). For a different view, see Lothar Haselberg, "Architectural Likenesses: Models and Plans of Architecture in Classical Antiquity," *Journal of Roman Archaeology* 10 (1997): 77–94. The controversy is compounded by the fact that archaeologists sometimes fail to distinguish between, on the one hand, *survey* drawings and models (the documentation of existing structures) and, on the other hand, *project* drawings and models (the design of buildings prior to their construction). Many extant small-scale architectural models from classical antiquity may have been copies of existing buildings, or they may have served for the presentation of new projects to the general public or to patrons, but it is unlikely that they may have been used for design purposes, as—until very recent times—models could not provide reliable measurements to construction workers. Unlike on-site, full-scale construction drawings, sometimes incised in marble walls (such as those recently discovered on the walls of the temple of Apollo at Didyma, Turkey), small-scale, proportional project drawings in classical antiquity are rare, and the few Haselberg reviews are in fact diagrams, not drawings. On post-Hellenistic and Roman use of correlated plans, elevations, and scaled models see Mark Wilson Jones, *Principles of Roman Architecture* (New Haven: Yale University Press, 2000), 50–56, 58–63.

24. Vitruvius, *De architectura*, 1.2.2. (Unless otherwise indicated, all citations of Vitruvius's Latin text follow the Loeb Classical Library edition: *Vitruvius on Architecture*, ed. and trans. Frank Granger [London: Heinemann; Cambridge, MA: Harvard University Press, 1931–1934].) See Coulton, *Ancient Greek Architects*, 68, and below, section 1.5. The discrepancy between Vitruvius's declaration of principles and the core of his theory throughout his treatise may be one of the many anachronisms ensuing from Vitruvius's well-known doxographic practice: his theory of architectural drawings in Book I may draw from more recent sources than his design theory in Books III and IV; these, in turn, would hark back to (and, incidentally, pass on to posterity) an older, pre-Hellenistic tradition. In the Renaissance, Vitruvius's *ichnographia* and *orthographia* were commonly interpreted as scaled plans and elevations, as per Alberti's (and Raphael's) definitions, and *scaenographia* as a perspectival view, or something similar to it; self-interest may have prompted Daniel Barbaro and Palladio to interpret Vitruvius's *scaenographia* (which they spelled "sciographia") as "profilo" (whereby they meant a scaled section drawing), thus making Vitruvius's theory, and Palladio's practice, perfectly consistent with one another. See *I dieci libri dell'Architettura di M. Vitruvio tradutti e commentati da Monsignor Barbaro* ... (Venice: Marcolini, 1556), 19; *M. Vitruvii Pollionis de Architectura Libri decem, cum Commentariis Danielis Barbari* ... (Venice: Francesco de Franceschi and Ioan Crugher, 1567), 18; *I dieci libri dell'Architettura di M. Vitruvio, tradotti e commentati da Mons. Daniel Barbaro ... da lui riveduti ed ampliati* (Venice: Francesco de Franceschi, 1584), 29–30. See also James S. Ackerman, *Origins, Imitation, Conventions* (Cambridge, MA: MIT Press, 2002), 224–225. On the meaning of Vitruvius's *scaenography*, and its relation to the knowledge of perspective in classical antiquity, see Pierre Gros, "The Theory and Practice of Perspective in Vitruvius's De Architectura," in Mario Carpo and Frédérique Lemerle, eds., *Perspective, Projections and Design* (London: Routledge, 2007), 5–17. Even the notion that Vitruvius's passage on architectural drawings in his first book may refer to plans and elevations proportionally drawn to scale is dubious, as it rests on the interpretation of just one word ("modice"). Recent translators of Vitruvius have somewhat generously translated "modice" as "drawn to scale" (see Vitruvius, *Ten Books ...*, ed. and trans. Ingrid D. Rowland and Thomas N. Howe [Cambridge: Cambridge University Press, 1999], 25), apparently on the grounds that "there is vast evidence of the use of scaled drawings in antiquity, even though

a craftsman with no training as a painter, and he disregards another version of the letter with an added coda on perspectival drawings (where the text attributed to Raphael recommends using perspective to show aspects that cannot be seen when buildings are only drawn and "measured architecturally.")

28. Guidobaldo Del Monte, *Planisphaeriorum universalium theoricae* (Pesaro, 1579), II, 58. See Filippo Camerota, "'The Eye of the Sun': Galileo and Pietro Accolti on Orthographic Projections," in Carpo and Lemerle, eds., *Perspective, Projections and Design*, 118n12.

29. Piero della Francesca, "De prospectiva pingendi," ca. 1490, MS 1576, Book 3, proposition 8, fol. 64r, Biblioteca Palatina, Parma.

30. The use of the expression "orthogonal projections" (or "orthographic projections") to indicate early modern plans and elevations (and/or side views or sections, as defined by Alberti and Raphael, respectively) is frequent among scholars, but is misleading as it implies a theory of parallel projections that did not exist at the time. On central and parallel, or pseudo-parallel, projections in the Renaissance, see Robin Evans's pioneering "Translations from Drawing to Building," *AA Files* 12 (1986): 3–18 (republished in Evans, *Translations from Drawing to Building and Other Essays* [London: Architectural Association Publications, 1997], 154–193); see also James Ackerman's review of the latter in *Design Book Review* 41/42 (2000): 65–67; Evans, "Architectural Projections," in Eve Blau and Edward Kaufman, eds., *Architecture and Its Image* (Montreal: Canadian Centre for Architecture, 1989), 19–35; Carpo and Lemerle, introduction to *Perspective, Projections and Design*, 1–5; and esp. Camerota, "'The Eye of the Sun'," 115–127. The expression "orthographic projections" seems to have first occurred in François D'Auguillon (Aguilonius), *Opticorum libri sex* (Antwerp, 1613), 683 (see Camerota, 118n11), where, however, it referred to shadows projected by the direct light of the sun (unlike a physical eye, the sun could be assimilated to a geometrical point located at infinity). Indeed, with the exception of Desargues's theory of the conics (1639), which had no immediate follow up, pre-Mongeian parallel projections were mostly discussed with regard to astronomy and to the theory of shadows (or to artificially distorted views, such as those generated by the lenses of the telescope): see Camerota, 125. For different takes on this controversial matter, see Alberto Pérez-Gómez and Louise Pelletier, *Architectural Representation and the Perspective Hinge* (Cambridge, MA: MIT Press, 1997), and Branko Mitrovic, "Leon

Battista Alberti and the Homogeneity of Space," *Journal of the Society of Architectural Historians* 63, no. 4 (2004): 424–439.

31. Leon Battista Alberti, *De re aedificatoria*, 2.1.4 (unless otherwise indicated, all citations of Alberti's Latin text follow *L'architettura [De re ædificatoria]*, ed. and trans. Giovanni Orlandi [Milan: Il Polifilo, 1966]): "Inter pictoris atque architecti perscriptionem hoc interest, quod ille prominentias ex tabula monstrare umbris et lineis et angulis comminutis elaborat, architectus spretis umbris prominentias istic ex fundamenti descriptione ponit, spatia vero et figuras frontis cuiusque et laterum alibi constantibus lineis atque veris angulis docet, uti qui sua velit non apparentibus putari visis, sed certis ratisque dimensionibus annotari." In the recent English translation the term "ratus" was translated not as "proportional," but as "according to certain calculated standards": see *On the Art of Building in Ten Books*, trans. Joseph Rykwert, Neil Leach, and Robert Tavernor (Cambridge, MA: MIT Press, 1988), 34. See, on the contrary, the translation and commentary in Alberti, *L'art d'édifier*, ed. and trans. Pierre Caye and Françoise Choay (Paris: Seuil, 2004), 100n9, pointing out the novelty of Alberti's "proportional" drawings. Alberti's requirement of "constantes lineae" in architectural design may not refer to generic "coherent lines," as hitherto translated, but more technically to "lines with constant value throughout," meaning uniform or consistent mensural value. This would in turn suggest, again, lines that should not be perspectivally foreshortened.

32. Alberti, *De re aedificatoria*, 1.1.2; 2.1.2; 2.1.4; 2.2.1; 2.3.1–3; 9.8.10; 9.9.5; 9.10.3–4; 9.11.1–2; 9.11.4–5. The same topic occurs, in a curiously different context, in *Momus*, one of Alberti's most enigmatic writings and almost coeval with *De re aedificatoria*. When Jupiter, apparently unsatisfied with the state of the world, ponders destroying it and constructing a new one from the ground up, Momus advises him that to best carry out his plan, thinking and action should be kept separate, and that the design of the new project ("descriptio") should be completed before construction starts. Leon Battista Alberti, *Momus*, ed. and trans. Virginia Brown and Sarah Knight (Cambridge, MA: Harvard University Press, 2003), 191.

33. The expressions Alberti uses to refer to project drawings and/or models (sometimes the difference is not clear) in *De re aedificatoria* are "lineamenta" and "perscriptio," 1.1.2: also recorded as "praescriptio" and "proscriptio" in some

manuscripts and in the *editio princeps*: see Orlandi, 21; "perscriptio," "pictura," modulus," and "exemplarium" (2.1.4); "modulus" (2.2.1 and 2.3.1–3); "exemplar" ("ad modulus diductum," 9.8.10); "exemplar" (9.9.5); "institutum" (9.10.3); "modulus" and "exemplar" (9.10.4); "ad lineas redigere," "perscripta," "modulus," and "exemplar" (9.10.11); "castigata lineamenta" (9.11.1).

34. *De re aedificatoria*, 9.11.1: "Dignitatem idcirco servasse consulti est; fidum consilium poscenti castigataque lineamenta praestitisse sat est." (to someone who asks for your work, it will suffice that you provide sound advice and clear drawings). My translation; *On the Art of Building*, 318, translates differently.

35. *De re aedificatoria*, 1.1.2; *On the Art of Building*, 7. As Alberti further remarks when he sums up his tenets of the architect's profession at the end of Book IX: "how often have I conceived of projects in the mind that seemed quite commendable at the time; but when I translated them into drawings, I found several errors in the very parts that delighted me most, and quite serious ones; once again, when I return to drawings, and measure the dimensions, I recognize and lament my carelessness; finally, when I pass from the drawings to the model, I sometimes notice further mistakes in the individual parts, even over the numbers." *De re aedificatoria*, 9.10.11; *On the Art of Building*, 317.

36. *De re aedificatoria*, 9.8.10; *On the Art of Building*, 313. On the need to seek external advice, see also two more passages in *De re aedificatoria*, 2.1.4; *On the Art of Building*, 34; and 9.11.6; *On the Art of Building*, 319.

37. *De re aedificatoria*, 2.3.1. My translation; *On the Art of Building*, 37, translates differently.

38. In Book II of *De re aedificatoria* Alberti advises against all changes over the course of construction, including "authorial" revisions, after the project has been finalized (2.1.2); the argument is restated in Book IX, when Alberti emphasizes that the designer should abstain from all intervention after construction has begun. Building should start only when the design is completed in all its aspects (2.1.4); finances and logistics should be planned is such a way that when building starts it may proceed speedily and steadily, without hindrances or shortages of materials (2.3.3); before building starts, everything should be checked many times and experts should be consulted (9.8.10); after the construction has begun, no changes nor hesitations are allowed, nor suspension of the works (9.9.5); construction in compliance with project and budget is not the architect's job but the workers':

"facere, quae usui commoda videantur, et quae posse pro instituto et fortunae ope fieri non dubites, non magis architecti est quam operarii fabri; sed praecogitasse ac mente iuditioque statuisse, quod omni ex parte perfectum atque absolutum futurum sit, eius unius est ingenii, quale quaerimus" (9.10.3). Architects should provide patrons with nothing more than sound advice and clear drawings ("fidum consilium" et "castigata lineamenta," 9.11.1: see above, note 34).

39. *De re aedificatoria*, 9.11.2; *On the Art of Building*, 318.

40. *De re aedificatoria*, 9.11.4; *On the Art of Building*, 318.

41. See below, chapter 2, note 41.

42. *De re aedificatoria*, 9.11.5; *On the Art of Building*, 319. In classical Latin, "auctor," from "augere," designates the agent that makes something grow, variously translated as instigator, originator, inventor, creator, or maker. Classical writers, poets, or historians were occasionally referred to as "auctores," mostly to indicate their role in the invention or introduction of a new literary tradition or genre, whereas the authorship of literary works was often attributed to the patrons or dedicatees who had encouraged, nurtured, or commissioned them: see Florence Dupont, "Comment devenir à Rome un poète bucolique," in Claude Calame and Roger Chartier, eds., *Identités d'auteurs dans l'Antiquité et la tradition européenne* (Grenoble: Jérome Millon, 2004), 171–175. The term "auctor" occurs 17 times in *De re aedificatoria*, in most cases to denote a literary source: see Hans-Karl Lücke, *Alberti Index* (Munich: Prestel, 1975), *ad vocem*. But Alberti also uses the term in the classical, etymological sense: thus, David and Solomon were the "authors" of the Temple of Jerusalem (2.3.4; *On the Art of Building*, 38), Themistocles of the walls of Athens (7.2.1; 192), and Hercules was the "author" of sport competitions (8.7.2; 269). In one instance, the term refers to an artisan (Callimachus was the "author" of the first Corinthian capital, 7.6.3; 201). The designation of the architect as an author occurs only twice: in Book II (where the judgment of the "author," i.e., the designer, is juxtaposed to the expertise of executors of a building, 2.1.1; *On the Art of Building*, 33) and in the passage cited here. In both cases, as he refers to the designer as the only "author" of a building, Alberti conflates the two meanings of the term: the architect is the originator, inventor, and creator of the building but at the same time, the architect's design becomes as authoritative as any "authorized" literary text: in Alberti's theory, a stable, permanent, authorial source not to be altered, changed, or in any way tampered with by others.

43. See above, note 33. Brunelleschi's first biographer, Antonio di Tuccio Manetti, insistently suggested that Brunelleschi considered drawing to be but a partial and inadequate notational tool: whenever Brunelleschi needed to convey a complete set of instructions for building, he resorted to three-dimensional models, and as Brunelleschi in most cases preferred not to reveal all the aspects of his design, he deliberately left such models incomplete. Antonio di Tuccio Manetti, *The Life of Brunelleschi*, ed. and trans. Howard Saalman and Catherine Enggass (University Park: Pennsylvania State University Press, 1970), 96, 116. See below, chapter 2, notes 35, 39, 40.
44. The design of an object may at times be a full-scale prototype of the object itself. This is not infrequent, particularly for smaller building components, and appears to have been customary in classical Greece: see Coulton, *Ancient Greek Architects*, 55, nn17–18.
45. Goodman, *Languages of Art*, 220.
46. Paul Zumthor, *Essai de poétique médiévale* (Paris: Seuil, 1972), 65–75; Bernard Cerquiglini, *Éloge de la variante: Histoire critique de la philologie* (Paris: Seuil, 1989), 57–79, 120; Herrad Spilling, ed., *La collaboration dans la production de l'écrit médiéval* (Paris: École des Chartes, 2003). Zumthor's and Cerquiglini's notion of variance refers to medieval poetry written in the French vernacular. On the "drifting" of early modern visual copies (before the "freezing" of printed images), see Wood, "Notation of Visual Information in the Earliest Archeological Scholarship." On the late medieval culture of textual variations (and Petrarch's early repudiation of it), see María Rosa Menocal, "The Flip Side," in Hans-Ulrich Gumbrecht and Michael Marrinan, eds., *Mapping Benjamin: The Work of Art in the Digital Age* (Stanford: Stanford University Press, 2003), 293–295, and Menocal, *Shards of Love: Exile and the Origins of the Lyric* (Durham: Duke University Press, 1994), esp. 176–181. See below, pages 115–119.
47. Commenting on the tortuous textual history of Poggio Bracciolini's *De varietate fortunae*, Anthony Grafton concludes: "[Poggio] created a new kind of text: a written work containing apparently firsthand reports on material facts, arguments, and evidence drawn from stones and inscriptions that were thousands of years old. And he suggested that this sort of scholarship, unlike the writing of narrative history, must rest on the work not of a single gifted and experienced individual, but of a community whose members collaborated with and corrected one another."

Grafton, *Leon Battista Alberti, Master Builder of the Italian Renaissance* (New York: Hill and Wang, 2000), 229. In the same book, Grafton also remarks that Alberti himself often kept revising and sending out drafts of his works, asking for comments and feedback from experts and friends: "When Alberti called on critics to 'emend' works of art—and asked Brunelleschi and other artists to 'emend' his own treatise—he was not repeating an empty commonplace. ... Even when he found himself frustrated by particular critical responses ... he never lost faith in the idea that a collaborative, social system of literary production would yield better results than a completely individualistic one." In doing so, "Alberti had in mind a particular literary system—the very one in which he himself struggled, during the 1430s, to attain intellectual mastery. This system was as collaborative in form as bound to the assumption that creativity is a social, not an individual process, as the system for commissioning art to which Alberti connected it" (138–139). In the case of his treatise *Della famiglia*, Alberti's revisions resulted in the circulation of multiple and different versions (although Grafton also refers to one "final version of the text"; 175). Françoise Choay has recently suggested that Alberti's writing of *De re aedificatoria* should be seen as a permanent work in progress, thus excluding on conceptual grounds the very same notion that a final version of the work could exist, let alone be printed: introduction to Alberti, *L'art d'édifier*, 18–19.

48. See below, pages 115–119.

49. This term was introduced in architectural discourse by Ingeborg Rocker, "Versioning: Evolving Architecture, Dissolving Identities," in *Architectural Design* 72, no. 5 (2002): 10–18.

50. Cerquiglini, *Éloge de la variante*, 22–23.

51. In Goodman's terms, the compliance class of architectural notations is often limited to one building, as in the case of a symphony played only once. *Languages of Art*, 219.

52. Even in more recent times, when they were customarily reproduced in small numbers, construction drawings were rarely printed; they were copied using technologies based on chemical contact, such as the cyanotype (or blueprint) and its successors.

53. Exception made, crucially, for the variability inherent in today's digitized design and production processes; and, accidentally, for the scale of today's digital fabrication tools, which tend to produce smaller objects than buildings (apart from contour crafting or recent work by Thomas Heatherwick, who has used a peculiar CNC milling machine to carve actual houses out of mounds of concrete bricks).

54. For Eisenman's pivotal role in the early, "Deleuzian" phase of the digital turn, see below, pages 85–89. On Eisenman's more recent criticism of visual culture in the digital age, and on the digitally induced reshaping of architectural indexicality, see esp. chapter 3, note 41.

55. Nelson Goodman is quite particular in setting out the technical requirements of full notationality: see *Languages of Art*, 132–171.

56. As, for example, the radius or diameter of the base of the column for the columnar system, the lowest fascia of the architrave for the entablature, a pin hole in the structural frame of a catapult, etc. This method is most thoroughly described in Books III and IV on the building of temples, but is applied throughout the treatise, including Book X on the construction of machines.

57. See notes 23–24 above.

58. Mario Carpo, "Drawing with Numbers: Geometry and Numeracy in Early-Modern Architectural Design," *Journal of the Society of Architectural Historians* 62, no. 4 (2003): 448–469.

59. This term is due to Alberto Sdegno, "E-architecture: L'architettura nell'epoca del computer," *Casabella* 691 (2001): 58–67.

60. Almost concurrently with Christoph Scheiner's pantograph (invented between 1603 and 1631), similar mechanical devices were being developed for architectural purposes, most famously Ottavio Revesi Bruti's Archisesto (1627), a machine for the automatic scaling of the proportions of the architectural orders. Revesi Bruti's Archisesto, however, did not work, and was seldom used. See Carpo, "Drawing with Numbers," 467–468.

61. Robin Evans, *The Projective Cast: Architecture and Its Three Geometries* (Cambridge, MA: MIT Press, 1995), 302–303, 305.

62. Evans, *The Projective Cast*, 334.

63. See the work of Mark Burry, esp. "Virtually Gaudí," in Neil Leach, David Turnbull, and Chris Williams, eds., *Digital Tectonics* (Chichester: Wiley-Academy, 2004), 22–33.

64. This is irrespective of statics, construction science, the strength of building materials, or many other reasons why a geometrically defined body may not be buildable.

65. Le Corbusier, *Urbanisme* (Paris: G. Crès, [1924]), v.

66. Nicholas Negroponte, *The Architecture Machine* (Cambridge, MA: MIT Press, 1970).

67. See below, section 3.3 and chapter 4.

68. For the same reason, the actual chronology and technical history of the CAD and animation software that many young architects started to use around that time will not be discussed here.

69. See Anthony Vidler, *Warped Space: Art, Architecture, and Anxiety in Modern Culture* (Cambridge, MA: MIT Press, 2000); esp. 219–235 ("Skins and Bones: Folded Forms from Leibniz to Lynn"); Carpo, review of the exhibition "Architectures non standard," Centre Pompidou, Paris, December 2003–March 2004, and of the exhibition catalog *Architectures non standard*, ed. Frédéric Migayrou and Zeynep Mennan, *Journal of the Society of Architectural Historians* 64, no. 2 (2005): 234–235.

70. One of the best accounts to date of Gehry's early digital processes is in Alex Marshall, "How to Make a Frank Gehry Building," *New York Times Magazine*, April 8, 2001. See also Coosje van Bruggen, *Frank O. Gehry, Guggenheim Museum Bilbao* (New York: Guggenheim Museum, 1997), 135–140; Bruce Lindsey, *Digital Gehry: Material Resistance, Digital Construction* (Basel: Birkhäuser, 2001), 42–47, 65–69, and Sdegno, "E-architecture," 2001, with further bibliography.

71. Most of Scheiner's *Pantographice, seu ars delineandi res quaslibet per parallelogrammum* (Rome: Grignani, 1631) is devoted to the better-known planar pantograph (which Scheiner called "epipedographic"). The "stereographic" pantograph is in fact a perspectograph: a machine for drawing in perspective. A more advanced version of the latter can also be used to reproduce images of three-dimensional objects on irregular or nonplanar surfaces, but Scheiner appears to have devised it as a tool for the mechanical drawing of anamorphic images (100; Scheiner does not use the term "anamorphic"). Both the "epipedograph" and the "stereograph" are illustrated on the title page (the latter in the basic version, not the anamorphic one).

72. See chapter 3, notes 26–27.

73. Greg Lynn, "Blobs (or Why Tectonics Is Square and Topology Is Groovy)," *ANY* 14 (May 1996): 58–62.

74. Mario Carpo, "L'architecture à l'ère du pli (Architecture in the Age of Pliancy)," *L'architecture d'aujourd'hui* 349 (2003): 98–104.

75. Theories inspired by the biological, or genetic, model of development of different phenotypes from a single genotype. As parametric functions generate families

of similar curves from a single notation, and digital design is often (deliberately or not) based on algorithmically generated variations, the analogy between generative digital scripts and DNA-like developments of form is self-evident, and it has been vastly investigated, both critically and creatively. The analogy remains nonetheless purely metaphorical; in spite of recent scientific advancements in the field, the mechanic and the organic still belong, for most practical purposes, to different kingdoms of nature.

76. Between March 10, 2000 and October 9, 2002 the NASDAQ Composite Index lost 78 percent of its value (5048 to 1114). See Carpo, "Post-Hype Digital Architecture: From Irrational Exuberance to Irrational Despondency," *Grey Room* 14 (2004): 102–115.

77. Sometimes also called mass customization and, in this book, differential reproduction (following Greg Lynn: see note 10 above).

78. This term appears to have been introduced by the publisher and technologist Tim O'Reilly, and it rose to prominence after the first Web 2.0 Conference held by O'Reilly Media in San Francisco on October 5–7, 2004.

79. See esp. chapter 3, note 73.

80. The expression harks back to Robin Evans's seminal essay "Translations from Drawing to Building" (see note 30 above).

81. Carpo, "Pattern Recognition," 44–58.

82. Erwin Panofsky, *Gothic Architecture and Scholasticism* (Latrobe, PA: Archabbey Press, 1951).

83. Richard Krautheimer, "Introduction to an 'Iconography of Mediaeval Architecture'," *Journal of the Warburg and Courtauld Institutes* 5 (1942): 1–33; reprinted in *Studies in Early Christian, Medieval, and Renaissance Art* (New York: New York University Press, 1969), 115–151; see in particular 117–127, nn 82–86.

84. The literal opposite of the "generic" object in a genus/species relationship is a "specific" object.

CHAPTER 2

Parts of this chapter are excerpted from, and developed at greater length in Carpo, "Alberti's Media Lab," in Mario Carpo and Frédérique Lemerle, eds., *Perspective, Projections and Design* (London: Routledge, 2007), 47–63, and "Introduction: The Reproducibility and Transmission of Technico-Scientific Illustrations in the Work of Alberti and in His Sources," in *Leon Battista Alberti's "Delineation of the City of Rome" ("Descriptio Vrbis Romæ")*, ed. Mario Carpo and Francesco Furlan (Tempe, AZ: Center for Medieval and Renaissance Texts and Studies, 2007), 3–18.

1. Mario Carpo, *Architecture in the Age of Printing* (Cambridge, MA: MIT Press, 2001), 16–22, 122–124, and footnotes. The diagram had long been a visual ersatz of choice: many scientific writers in the classical tradition, first and foremost Euclid, had instructed the reader to draw diagrams, but such drawings were seldom meant to convey any additional or independent information: in many cases, all the information necessary to compose the diagrams was included in the text, where the meaning was carried exclusively by letters and numbers. On the transmission of Euclid's diagrams, see J. V. Field, "Piero della Francesca's Perspective Treatise," in Lyle Massey, ed., *The Treatise on Perspective: Published and Unpublished* (New Haven: Yale University Press; Washington: National Gallery of Art, 2003), 73–74. For a different view, see Reviel Netz, *The Shaping of Deduction in Greek Mathematics* (Cambridge: Cambridge University Press, 1999).
2. Of which the best known may be the description of the architectural moldings as combinations of the signs of the letters of the alphabet in the seventh book of *De re aedificatoria* (Alberti, *De re aedificatoria*, 7.7.10; *On the Art of Building in Ten Books*, trans. Joseph Rykwert, Neil Leach, and Robert Tavernor [Cambridge, MA: MIT Press, 1988], 574–575).
3. One work of Alberti's, the *Ludi rerum mathematicarum*, appears to have been illustrated, but most likely due to special and extenuating circumstances: see Francesco Furlan, introduction to *Leon Battista Alberti's "Delineation of the City of Rome,"* 23. On the apparent paradox of Alberti's outspoken refusal of illustrations in *De re aedificatoria*, a treatise for the most part devoted to a new theory and practice of architectural drawings, see Carpo, "Alberti's Media Lab," 49–51.
4. See Carpo, "Ecphrasis géographique et culture visuelle à l'aube de la révolution typographique," in Leon Battista Alberti, *Descriptio urbis Romae*, ed. Martine Furno and Mario Carpo (Geneva: Droz, 2000), 65–97; Carpo, "Introduction: The Reproducibility and Transmission of Technico-Scientific Illustrations in the Work of Alberti and in His Sources," in *Leon Battista Alberti's "Delineation of the City of Rome,"* 3–18.
5. See Bruno Queysanne, *Alberti et Raphaël, Descriptio urbis Romae, ou comment faire le portrait de Rome* (Grenoble and Lyon: École d'architecture de Grenoble and Plan Fixe, 2000; 2nd edn., Paris: Éditions de la Villette, 2002).

6. Leon Battista Alberti, *On Painting and On Sculpture: The Latin texts of De pictura and De statua*, ed. and trans. Cecil Grayson (London: Phaidon, 1972), 128–130.
7. J. Lennart Berggren and Alexander Jones, *Ptolemy's Geography: An Annotated Translation of the Theoretical Chapters* (Princeton: Princeton University Press, 2000), 4–5.
8. Alberti, *On Painting and On Sculpture*, 55.
9. Ibid., 49.
10. Ibid., 69.
11. Alberti wrote two versions of *De pictura*, one in Latin and one in Italian, with minor but meaningful differences between the two. See Alberti, *De pictura*, in *Opere volgari*, ed. Cecil Grayson (Bari: Laterza, 1973), 3: 54–55 (collated Italian and Latin versions); Alberti, *On Painting and On Sculpture*, 68–69 (Latin text and English translation). On the dating of *De pictura* (1435–1436) see Alberti, *Opere volgari*, 3: 305. Neither version of the treatise was illustrated, but illustrations of Alberti's window can be found in later editions of *De pictura*, and in many treatises on perspective that followed—most famously, in Dürer's *Unterweisung der Messung* (*Manual of Measurements*, 1525 and 1538: see here figure 2.4).
12. The dedicatory letter to Brunelleschi is translated into English in Alberti, *On Painting and Sculpture*, 33.
13. Friedrich A. Kittler, "Perspective and the Book," *Grey Room* 5 (2001): 44. As a pixel is an image unit described by a single value, Alberti's "analog" grid cannot in fact be equated to a pixellated raster, unless its graphic resolution is pushed to the geometrical limit.
14. Ibid. See also the notion of "pantometry," related to the early modern pursuit of visual precision, in Alfred W. Crosby, *The Measure of Reality: Quantification and Western Society, 1250–1600* (Cambridge: Cambridge University Press, 1997).
15. Alberti also experimented with other optical devices, similar to today's show boxes or slide viewers (with a picture on a semitransparent surface on one side of the box and a peephole at the opposite side), intended to recreate the ideal conditions for the observation of perspectival images from a single and fixed point of vision, thus reenacting in three dimensions the geometrical constructions described in *De pictura*. See Carpo, "Alberti's Media Lab," 55–56.
16. *De re aedificatoria*, 9.7.4–7; *On the Art of Building*, 310; *L'architecture et art de bien bastir*, trans. Jean Martin (Paris: Kerver, 1553), 136v–137. In James Leoni's first

English translation (from Cosimo Bartoli's Italian), these identical "statues, pictures, and ornaments" are described as "twins." *The architecture of Leon Battista Alberti in ten books* (1726; London: R. Alfray, 1755), 201. Citation is to the 1755 edition. Neither Alberti nor Leoni had anticipated contemporary genetic cloning.

17. Unrelated to the present topic, "Powers of the Facsimile: A Turing Test on Science and Literature" is the title of a recent essay by Bruno Latour on contemporary writer Richard Powers, published in Stephen J. Burn and Peter Dempsey, eds., *Intersections: Essays on Richard Powers* (Urbana-Champaign, IL: Dalkey Archive Press, 2008), 263–292.

18. "This shrine is the like of the Sepulcher in Jerusalem."

19. See Carpo, "Alberti's Media Lab," 56–59, with further bibliography. See also Jan Pieper, "The Garden of the Holy Sepulchre in Görlitz," *Daidalos* 58 (December 1995): 38–43; Pieper, "Jerusalemskirchen: Mittelalterliche Kleinarchitekturen nach dem Modell des Heiligen Graben," *Bauwelt* 80, no. 3 (January 1989): 82–101; Anke Naujokat, *Pax et concordia: Das Heilige Grab von Leon Battista Alberti als Memorialarchitektur des Florentiner Unionskonzils (1439–1443)* (Freiburg im Breisgau: Rombach Verlag, 2006); and Naujokat, *Ad instar iherosolimitani sepulchri. Gestalt und Bedeutung des Florentiner Heiliggrabtempietto von L. B. Alberti* (Ph.D. dissertation, University of Aachen, 2008).

20. See the famous (and controversial) notion of a "nonvisual" form of imitation in the Middle Ages in Richard Krautheimer, "Introduction to an 'Iconography of Mediaeval Architecture'," *Journal of the Warburg and Courtauld Institutes* 5 (1942): 1–33; reprinted in *Studies in Early Christian, Medieval, and Renaissance Art* (New York: New York University Press, 1969), 115–151; see in particular 117–127, nn 82–86.

21. See Cecil Grayson, "Alberti's Works in Painting and Sculpture," in Alberti, *On Painting and On Sculpture*, 143–154. According to Vasari, whose bias against Alberti is often patent, Alberti painted little, and his few extant works as a painter were bad: see Giorgio Vasari, *Le Vite ... nelle redazioni del 1550 e 1568*, ed. Rosanna Bettarini and Paola Barocchi (Florence: Sansoni, 1971), 3: 288–289.

22. Ibid., 3: 289.

23. *Vita di Leon Battista Alberti, di autore anonimo, con a fronte il volgarizzamento del dott. Anicio Bonucci*, in *Opere volgari di Leon Battista Alberti*, ed. Anicio Bonucci (Florence: Tipografia Galileiana, 1843–1849), cii; Riccardo Fubini and Anna Menci

Gallorini, "L'autobiografia di Leon Battista Alberti. Studio ed edizione," *Rinascimento*, ser. 2, 12 (1972): 73; Grayson, "Alberti's Works in Painting and Sculpture," 143 (for a different English translation of the same passage).

24. Alberti, *De pictura*, in *Opere volgari*, ed. Grayson, 3: 106–107; *On Painting*, 107.

25. Alberti missed the print revolution by only a handful of years. In his entire corpus there is but one mention of the invention of printing, at the opening of *De cifris*—ironically, a manual of cryptography—where Alberti reports a conversation that took place in Rome around 1466. See Carpo, *Architecture in the Age of Printing*, 118–119. In his preface to the *editio princeps* of *De re aedificatoria* (Florence: Niccolò Lorenzo Alamanno, 1485), Politian reports that not long before his death Alberti was preparing to "publish" his treatise on architecture ("editurus in lucem": cf. also *De re aedificatoria*, 3–4). As Françoise Choay recently pointed out, this passage may not mean that Alberti was planning for publication in print: see her introduction to Leon Battista Alberti, *L'art d'édifier*, ed. Pierre Caye and Françoise Choay (Paris: Seuil, 2004), 18–19.

26. See for example Martin Kemp, *The Science of Art: Optical Themes in Western Art from Brunelleschi to Seurat* (New Haven: Yale University Press, 1990); Barbara Maria Stafford and Frances Terpak, eds., *Devices of Wonder: From the World in a Box to Images on a Screen* (Los Angeles: Getty Research Institute, 2001); but also David Hockney, *Secret Knowledge: Rediscovering the Lost Techniques of the Old Masters* (New York: Viking Studio, 2001), and the controversies that followed.

27. Nelson Goodman, *Languages of Art: An Approach to a Theory of Symbols* (Indianapolis: Bobbs-Merrill, 1968; 2nd edn. 1976), 122. Citations are to the second edition.

28. For the history of the construction of the dome, see in particular Antonio Manetti, *Vita di Filippo Brunelleschi, preceduta da La novella del grasso*, ed. Domenico de Robertis and Giuliano Tanturli (Milan: Il Polifilo, 1976); Manetti, *The Life of Brunelleschi*, ed. and trans. Howard Saalman and Catherine Enggass (University Park: Pennsylvania State University Press, 1970); Howard Saalman, *Filippo Brunelleschi: The Cupola of Santa Maria del Fiore* (London: A. Zwemmer, 1980).

29. Saalman, in Manetti, *Life of Brunelleschi*, 139.

30. Brunelleschi and Ghiberti were continually reaffirmed as "capomaestri" of the cupola until June 1436. Brunelleschi's salary was indeed increased above Ghiberti's in 1426, whereas Ghiberti claims that he stayed with the same pay as Brunelleschi till the end. Saalman, in Manetti, *Life of Brunelleschi*, 139; Tanturli, in Manetti, *Vita*, 90–97.

31. Saalman, in Manetti, *Life of Brunelleschi*, 141n108; Tanturli, in Manetti, *Vita*, 97; Vasari, *Vite*, 3: 173–175.
32. *La novella del grasso*, in Manetti, *Vita*, 1–45.
33. Natalie Zemon Davis, *The Return of Martin Guerre* (Cambridge, MA: Harvard University Press, 1983).
34. For a discussion of some extant models of the dome and the lantern, sometimes attributed to Brunelleschi, see Massimo Scolari, entries 261–263 in Henry A. Millon and Vittorio Magnago Lampugnani, eds., *The Renaissance from Brunelleschi to Michelangelo: The Representation of Architecture* (London: Thames and Hudson, 1994), 584–586.
35. Manetti, *Life of Brunelleschi*, 76. Brunelleschi's role in the submission of the 1420 document is not fully known. See also Manetti's conclusions, ibid., 116: "After he had compiled some years of experience in the practice of architecture, Filippo's tendency—or, to put it better, his custom—concerning the models he made for the buildings that were required and which he built was to indicate little about the symmetrical elements. He took care only to have the principal walls built and to show the relationship of certain members without the ornaments. … For this reason the models for the building of Santa Maria degli Angeli and Santo Spirito were made in that way. He did not want to make a model for the Barbadori House or for the Parte Guelfa and instead worked with drawings alone, step by step telling the stonecutters and bricklayers what to do." (Saalman and Enggass's "symmetrical elements," cited above, translate Manetti's "simitrie," which at the time meant "proportions," in the classical and Vitruvian sense of the term.) On Brunelleschi's model for the lantern, which he left deliberately incomplete lest "whoever would make the model should discover his every secret," see note 40 below.
36. Frank D. Prager, "A Manuscript of Taccola, Quoting Brunelleschi, on Problems of Inventors and Builders," *Proceedings of the American Philosophical Society* 112, no. 3 (June 21, 1968): 139–141.
37. Manetti, *Life of Brunelleschi*, 94.
38. Vasari, *Vite*, 3: 159.
39. The most famous of these mistakes occurred in the building of the portico of the Ospedale degli Innocenti, for which Manetti says Brunelleschi had provided a

scaled drawing but no wood model, wrongly assuming that drawings would have been sufficient; the same careless master builder (according to Saalman, Francesco della Luna) would have also altered Brunelleschi's plans for the Palazzo di Parte Guelfa (*Life of Brunelleschi*, 97, 101).

40. Brunelleschi's model for the lantern was approved in 1436, but with some strings attached (Saalman, *Filippo Brunelleschi*, 139–142). On Brunelleschi's request in his testament that works on the lantern should proceed in full compliance with his "model" and "writings" after his death, see Vasari, *Vite*, 3: 179. Quite to the contrary, Manetti emphasizes that in many other cases Brunelleschi deliberately provided only incomplete or rough models, or no model at all, implying that Brunelleschi's "models" would have included richer and more complete instructions than his drawings, or that drawings were more difficult to interpret. Manetti goes on to suggest that Brunelleschi followed the same secretive tradition when he furnished the model for the lantern ("if the elements were not made so carefully in the model he did not care, but it seemed he was concerned that whoever would make the model should not discover his every secret, expecting to make all things precisely and well as they followed bit by bit in the actual building.") In this instance, Manetti concludes, Brunelleschi did not live to provide the workers with further instructions, and this explains some "incompleteness" in the lantern as built. *Life of Brunelleschi*, 116.

41. See Alberti's famous letter to Matteo de' Pasti, dated November 18, 1454, sent from Rome to Rimini. Alberti insists that his original "model" and "drawing" for the church of San Francesco should be followed without incompetent changes, and he even tries to explain and justify some of his design choices. The drawing of the scroll in the letter is an emphatic restatement of a detail already shown in the model; the same for the proportions of some pillars which, Alberti insists, can be seen in "my model": Alberti warns that if one alters those proportions, "si discorda tutta quella musica" (you bring discord to all that music). Alberti also insists on the necessity of following other parts of his design, particularly for the vaulting and oculi (see his somewhat exasperated conclusion: "and this I told you to show from where the truth arises"). We also infer from a letter of the same Matteo de' Pasti to Sigismondo Malatesta (December 17, 1454) that Alberti had provided a drawing of the facade and of a capital, alongside a model in wood. One comes to the almost inevitable conclusion that Alberti's notations (in drawings, models, and letters) must have been difficult to interpret and/or impossible to enforce. For an English translation of the letter, see Robert Tavernor, *On Alberti and the Art of Building* (New Haven: Yale University Press, 1998), 60–63, n 64.

42. Manetti, *Life of Brunelleschi*, 119.
43. On the chronology (and attribution) of Manetti's *Vita*, see Saalman, in Manetti, *Life of Brunelleschi*, 10–11; on Manetti's "anti-Albertian bias," ibid., 30. On Manetti's Savonarolan leanings, see Tanturli, introduction to Manetti, *Vita*, xxxvii; Antonio Manetti, *Vita di Filippo Brunelleschi*, ed. C. Perrone (Rome: Salerno, 1992), 30.
44. John Ruskin, *The Stones of Venice* (London: Smith, Elder, 1851–1853), III, IV, 35, 194.
45. Goodman, *Languages of Art*, 218–221.
46. This does not mean that that building would necessarily stand up; the sole purpose of notations, as discussed in this context, is the geometrical definition of an object.

CHAPTER 3

Parts of this chapter are adapted from previously published materials: Section 3.1 is adapted from Carpo, "Ten Years of Folding," introductory essay to Greg Lynn, ed., *Folding in Architecture* (London: Wiley-Academy, 2004), 6–14 (reprint of a special issue [AD Profile 102] of *Architectural Design* 63, nos. 3–4 [1993]); and from "L'architecture à l'ère du pli," *Architecture d'Aujourd'hui* 349 (2003): 98–104. Section 3.2 is adapted from Carpo, "Non-standard Morality," in Anthony Vidler, ed., *Architecture between Spectacle and Use* (Williamstown, MA: Clark Art Institute; New Haven: Yale University Press, 2008), 127–142; and from "Tempest in a Tea-pot," *Log* 6 (2005): 99–106. Section 3.3 is adapted from conference papers read at the Southern California Institute of Architecture, Los Angeles, the Institute of Fine Arts, New York, and the Yale School of Architecture in the fall of 2007 and the spring of 2008, and partly published or abridged in Carpo, "Revolutions: Some New Technologies in Search of an Author," *Log* 15 (Winter 2009): 49–54, and "Revolución 2.0: El fin de la autoría humanista," *Arquitectura Viva* 124 (2009), 19–26.

1. Letter to Pauline Trevelyan, September 1854, in *The Works of John Ruskin*, ed. E. T. Cook and A. Wedderburn, vol. 36 (London: George Allen, 1909), 175.
2. Heinrich Wölfflin, *Kunstgeschichtliche Grundbegriffe* (1915); *Principles of Art History*, trans. M. D. Hottinger from the 7th rev. edn., 1929 (London: G. Bell and Sons, 1932), 230–235 ("Periodicity of Development," and "The Problem of Recom-

mencement"); on the "spiral movement," 234. See also Michael Podro, *The Critical Historians of Art* (New Haven: Yale University Press, 1982), 140.

3. Wolfgang Köhler, *Gestalt Psychology* (New York: Liveright, 1929).
4. Luis Fernández-Galiano has compared the "sharp folds of the F-117 Nighthawk Lockheed's stealth fighter" and the "undulating profile" of the later B-2 stealth fighter made by Northrop Grumman, considering the former as representative of the "fractured forms of deconstructivism that initiated the nineties under the wings of Derrida," and the latter as representative "of the warped volumes of the formless current that are wrapping up the decade, referring back to Deleuze or Bataille." "Split-screen: La décennie numérique," *Architecture d'Aujourd'hui* 325 (December 1999): 30. Oddly, the technical specifications—aerodynamics and the avoidance of radar detection—would have been the same for both fighter planes.
5. See Giuseppa Di Cristina, "The Topological Tendency in Architecture," in Di Cristina, ed., *Architecture and Science* (London: Wiley-Academy, 2001), 6–14.
6. Gilles Deleuze, *Le pli: Leibniz et le baroque* (Paris: Éditions de Minuit, 1988); *The Fold: Leibniz and the Baroque*, trans. Tom Conley (Minneapolis: University of Minnesota Press, 1993).
7. Peter Eisenman, "Unfolding Events: Frankfurt Rebstock and the Possibility of a New Urbanism," in Eisenman Architects, Albert Speer and Partners, and Hanna/Olin, *Unfolding Frankfurt* (Berlin: Ernst und Sohn, 1991), 8–18; Eisenman, "Visions' Unfolding: Architecture in the Age of Electronic Media," *Domus* 734 (January 1992): 17–24, reprinted in Jonathan Crary and Sanford Kwinter, eds., *Incorporations* (New York: Zone Books, 1992); Eisenman, "Folding in Time: The Singularity of Rebstock," in Lynn, ed., *Folding in Architecture*, 22–26 (all now reprinted in Eisenman, *Written into the Void: Selected Writings 1990–2004*, ed. Jeffrey Kipnis [New Haven: Yale University Press, 2007], 12–18, 34–41, 25–33).
8. Greg Lynn, ed., "Folding in Architecture," special issue (AD Profile 102), *Architectural Design* 63, nos. 3–4 (1993).
9. Gilles Deleuze, "The Pleats of Matter," trans. T. Conley, in Lynn, ed., *Folding in Architecture*, 17–21.
10. William Hogarth, *The Analysis of Beauty. Written with a view of fixing the fluctuating Ideas of Taste* (London: J. Reeves, 1753). Hogarth's *Analysis of Beauty* is cited by more recent writers on what is today sometimes called the digital baroque: see Patrik Schumacher, "Arguing for Elegance," in Ali Rahim and Hina Jamelle, eds.,

"Elegance," special issue (AD Profile 185), *Architectural Design* 77, no. 1 (2007): 36. For a more detailed genealogy of the serpentine line, from its origins in the theory of the picturesque to contemporary digital form making, see Lars Spuybroek, "The Aesthetics of Variation," in *The Architecture of Continuity* (Rotterdam: V2_Publishing, 2008), 245–263.

11. Deleuze, *Le pli*, 20–27.
12. Eisenman, "Unfolding Events," 14.
13. Greg Lynn, "Architectural Curvilinearity: The Folded, the Pliant and the Supple," in Lynn, ed., *Folding in Architecture*, 8–15. See esp. 13 on "the catastrophe diagram used by Eisenman in the Rebstock Park project, ... by Kipnis in the Briey project, and Shirdel in the Nara Convention Hall."
14. "[Deleuze] argues that in the mathematical studies of variation, the notion of object is change. This new object for Deleuze is no longer concerned with the framing of space, but rather a temporal modulation that implies a continual variation of matter. ... No longer is an object defined by an essential form. He calls this idea of an object, an 'object event.' The idea of event is critical to the discussion of singularity. Event proposes a different kind of time which is outside of narrative time or dialectical time." Eisenman, "Folding in Time," 24.
15. "These typologies, introduced into the system of the Fold, allow the Fold to reveal itself; the folding apparatus is invisible, purely a conceptual drawing, until it is activated by something cast into it." Eisenman, "Unfolding Events," 16.
16. For an overview of this discussion in essays by Michael Speaks, Greg Lynn, Jeffrey Kipnis, and Brian Massumi, see Di Cristina, "The Topological Tendency in Architecture," 6–14, esp. 10 and nn15–18; Michael Speaks, "It's Out There ... The Formal Limits of the American Avant-garde," in Stephen Perrella, ed., "Hypersurface Architecture," special issue (AD Profile 133), *Architectural Design* 68, nos. 5–6 (1998): 26–31, esp. 29: "Why does [Lynn's] architecture not move? ... Why does his architecture stop moving when it is no longer design technique and becomes architecture?"
17. Peter Eisenman, "Alteka Office Building," in Lynn, ed., *Folding in Architecture*, 28.
18. "Folding is only one of perhaps many strategies for dislocating vision." Eisenman, "Visions' Unfolding," 24.
19. Lynn, "Architectural Curvilinearity," 8.

20. Frank Gehry and Philip Johnson, "Lewis Residence, Cleveland, Ohio," in Lynn, ed., *Folding in Architecture*, 69.

21. Lynn, "Architectural Curvilinearity," 12; Jeffrey Kipnis, "Towards a New Architecture," in Lynn, ed., *Folding in Architecture*, 47.

22. Stephen Perrella, "Interview with Mark Dippe: Terminator 2," in Lynn, ed., *Folding in Architecture*, 93.

23. "Shoei Yoh, Odawara Municipal Sports Complex," in Lynn, ed., *Folding in Architecture*, 79. See also Lynn, "Classicism and Vitality," in Anthony Iannacci, ed., *Shoei Yoh* (Milan: L'Arca Edizioni, 1997), 15: "In all of these [Shoei Yoh's] projects there is a response to the shift in the economies and techniques of construction from one of assembly-line production of a standard to the assembly-like production of a series of singular units. These projects articulate an approach to standardization and repetition that combines a generic system of construction with slight variations of each member. This attribute is reminiscent of historic methods of craftsmanship where every element could be generic in some regard while given a distinct identity in each instance. ... Through both manual construction and industrial fabrication [these projects] exploit the economy of what is often referred to as 'custom assembly-line production.'"

24. A comprehensive history of splines is still to be written—from the first mechanical splines used in some specialized crafts, such as ship hull making, to the flexible rubber or metal strip used until recently for analog drafting, to the mathematical functions devised by Pierre Bézier, Paul de Casteljau, and others in the automotive industry (circa 1959–1962), which are at the basis of most contemporary software for the calculation of continuous lines and curved surfaces. See Spuybroek, "Textile Tectonics," in *The Architecture of Continuity*, 231. Recent research by Bernhard Siegert (Bauhaus-Universität Weimar) on early modern ship design, presented at several conferences, is unpublished at the time of writing.

25. See Edmund Burke, *A Philosophical Enquiry into the Origin of Our Ideas of the Sublime and the Beautiful* (1757); William Gilpin, *Observations [...] Relative Chiefly to Picturesque Beauty* (1782) and *Three Essays: On Picturesque Beauty; On Picturesque Travel; and On Sketching Landscape: To Which Is Added a Poem on Landscape Painting* (1792). In mathematical terms, the quality of smoothness of a line or surface is defined by the function that designates the angular coefficients of the tangents to each point

of it (that is, by the first derivative of the function that describes the original line or surface).

26. Bernard Cache, "Objectile: The Pursuit of Philosophy by Other Means," in Stephen Perrella, ed., "Hypersurface Architecture II," special issue (AD Profile 141), *Architectural Design* 69, nos. 9–10 (1999): 67.
27. For centuries, architects had been drawing with algebra, but now, "CAD software enables architects to draw and sketch using calculus." Greg Lynn, *Animate Form* (New York: Princeton Architectural Press, 1999), 16–18.
28. Deleuze, *Le pli*, 26.
29. Ibid., 20–27.
30. Bernard Cache, *Earth Moves: The Furnishing of Territories*, trans. Anne Boyman, ed. Michael Speaks (Cambridge, MA: MIT Press, 1995), iii. In French, *Terre meuble* (Orléans: Éditions HYX 1997).
31. The official date of birth of architectural blobs (i.e., of architectural blobs defined as such) appears to be May 1996. See Greg Lynn, "Blobs (or Why Tectonics Is Square and Topology Is Groovy)," *ANY* 14 (May 1996): 58–62. For a survey of blob developments in the late 1990s, see Peter Cachola Schmal, ed., *Digital Real. Blobmeister: Erste gebaute Projecte* (Basel: Birkhäuser, 2001).
32. Sigfried Giedion, *Mechanization Takes Command: A Contribution to Anonymous History* (1948; repr., New York: Norton, 1969), 87, 176.
33. The expression is due to Mauro F. Guillén, *The Taylorized Beauty of the Mechanical: Scientific Management and the Rise of Modernist Architecture* (Princeton: Princeton University Press, 2006).
34. Statistically, the constancy of the results is proportional to the number of samples, and increases if the more deviant samples are averaged out.
35. *Fortune* noted at the time that "the blending of whiskies from different batches to produce a consistent taste is one of the achievements of the Seagram organization." See Nicholas Faith, *The Bronfmans* (New York: St. Martin's Press, 2006), 100–104, for further references and bibliography.
36. I am grateful to Clarke Magruder of Georgia Tech, Atlanta, for bringing this story to my attention. Coca-Cola's new fountain dispenser will deliver directly to the customer "unprecedented beverage variety to suit any consumer taste," thus "celebrating the idea that consumers can truly have their say ... with choices tailored completely for them." ("The Coca-Cola Company Invites Consumers to Experi-

ence 'Freestyle.' New Proprietary Fountain Dispenser Gets a Brand Name," The Coca-Cola Company, news release, April 28, 2009). On a similar note, the award-winning German company Mymuesli.com, founded in 2007, offers an online shopping service that lets consumers create their own breakfast cereal from a list of 70 ingredients (web site accessed July 26, 2009). The technology used to mix the ingredients on demand is not stated. For small volumes the orders, which are submitted via scroll-down menus, could easily be manually executed. The individual recipes can be stored online for reordering, and packages are delivered by mail. Featured story in *DB, Das Magazin der Deutschen Ban* 7 (July 2009): 54–55.

37. See Gilles Deleuze, *Différence et répétition* (Paris: Presses Universitaires de France, 1968), esp. chap. 4, "Synthèse idéale de la différence," 218–235; for Deleuze's *Le pli*, see note 6 above.

38. From different vantage points, technologists and observers of the present visual environment are coming to similar conclusions. See Wendy Hui Kyong Chun, "On Software, or the Persistence of Visual Knowledge," *Grey Room* 18 (Winter 2005): 27–47.

39. More precisely, a tea and coffee service. See Greg Lynn, "Variations calculées," in Frédéric Migayrou and Zeynep Mennan, eds., *Architectures non standard* (Paris: éditions du Centre Pompidou, 2003), 91. According to commercial information furnished by Alessi SpA, the original project included fifty thousand variations, of which ninety-nine were made in addition to the author's three copies.

40. See Objectile (Patrick Beaucé and Bernard Cache), "Vers une architecture associative," in Migayrou and Mennan, eds., *Architectures non standard*, 138–139.

41. See Peter Eisenman, "Digital Scrambler, from Index to Codex," in Elijah Huge and Stephanie Turke, eds., "Building Codes," special issue, *Perspecta* 35 (2004): 40–53; reprinted in Eisenman, *Written into the Void*, 133–150: see esp. 147 (on Eisenman's project for the City of Culture of Galicia, Spain, 1999): "At Santiago, the indices become scrambled by a series of deformation and flow lines extrapolated from the original tartan grid. These lines represent the activity of a new digital—as opposed to analogic—code: a code that scrambles the prior notations. ... The generator of forms is now a torquing digital vector, a scrambler of the superposed grids. ... The resultant matrix is no longer an index of this activity. Because of the scrambler it cannot be traced back to some origin. There is no longer a linear narrative or legibility in presence." See also Carpo, "Notes on the End of the Index," in Andreas Beyer, Matteo Burioni, and Johannes Grave, eds., *Das Auge der Architektur: Zur Frage der Bildlichkeit in der Baukunst* (Munich: Wilhelm Fink Verlag, forthcoming); and the recent revival of the discussion on indexes and icons between Charles Jencks and Peter Eisenman (Eisenman, "Duck Soup," *Log* 7 [2006]: 139–141; Jencks and Carpo, "Letters," *Log* 9 [Winter/Spring 2007]: 7–12); Carpo, review of Charles Jencks, *The Iconic Building* (Winter/Spring 2005), in *Architecture d'Aujourd'hui* 368 (2007): 4–5.

42. On architectural notations (and blueprints in particular) as matrixes in an ideally mechanical, indexical process, see above, sections 1.4 and 2.5.

43. See Mario Carpo, "Pattern Recognition," in *Metamorph*, 3:44–58, and above, pages 8–11, 45–47.

44. See in particular the exhibition "Architectures non standard," curated by Frédéric Migayrou and Zeynep Mennan, Paris, Centre Pompidou, December 10, 2003–March 1, 2004; and Migayrou and Mennan, eds., *Architectures non standard*.

45. See Mario Carpo, review of the exhibition "Architectures non standard," and of Migayrou and Mennan, eds., *Architectures non standard*, *Journal of the Society of Architectural Historians* 64, no. 2 (2005): 234–235.

46. See, for example, the range of products and services, including proprietary software, offered by Gehry Technologies, based in Los Angeles (<http://gehrytechnologies.com>), the commercial operations of Bernard Cache's firm Objectile (<http://objectilediffusion.com>), or the various inclusive solutions for "low-volume manufacturing" featured by the Belgian company Materialise (<http://materialise.com>), including the .MGX design line (a seamless provider of CAD-CAM on demand, from proprietary design software to several 3D printing technologies, which can be remotely controlled by each individual customer, and which has proven popular with some well-known designers of high-tech furniture). All web sites accessed June 20, 2009. "Low-volume manufacturing" can be seen as a commercial approximation of the principles of nonstandard seriality, which, mathematically, could be defined as "zero-volume manufacturing." Likewise, several solutions for "print-on-demand" book publishing, which already exist on the market, offer products that are neither autographic manuscripts nor actual print runs, but something in between (which is already creating a host of whimsical problems for librarians, literary agents, and copyright offices). Many mainstream companies around the world have already quietly adopted integrated

CAD-CAM technologies for a variety of niche markets, from prefabricated building components to stone cutting to medical devices.

47. See Stanley M. Davis, *Future Perfect* (Reading, MA: Addison-Wesley, 1987), where the expression "mass customization" seems to have first occurred; and Joseph B. Pine, *Mass Customization: The New Frontier in Business Competition*, foreword by Stanley M. Davis (Boston: Harvard Business School Press, 1993). The term was brought to the attention of architects and designers by William J. Mitchell in the late 1990s: see his "Antitectonics: The Poetics of Virtuality," in John Beckmann, ed., *The Virtual Dimension: Architecture, Representations, and Crash Culture* (New York: Princeton Architectural Press, 1998), 205–217, esp. 210–212 on "Craft/Cad/Cam"; and *E-topia: "Urban Life, Jim, but Not as We Know It"* (Cambridge, MA: MIT Press, 1999), 150–152. See also Dagmar Steffen, "Produire en masse pour chacun," *Architecture d'Aujourd'hui* 353 (July-August 2004): 102–107.

48. The tables were produced by Cache's firm, Objectile, and those displayed at the exhibition are now in the permanent collection of the National Modern Art Museum at the Centre Pompidou in Paris. See "Tables non standard," Galerie Natalie Seroussi, Paris, May-June 2005. See also the uncredited review of the exhibition in *Architecture d'Aujourd'hui* 358 (May-June 2005): 38.

49. See Manuel De Landa, "Material Complexity," in Neil Leach, David Turnbull, and Chris Williams, eds., *Digital Tectonics* (Chichester: Wiley-Academy, 2004), 14–22, esp. 20–21.

50. See the recent work of Gramazio & Kohler at the ETH in Zurich, in particular the facade of the Gantenbein winery in Fläsch, in collaboration with Bearth & Deplazes, or the installation at the Swiss Pavilion at the 11th Biennale d'Architettura in Venice (2008). Fabio Gramazio and Matthias Kohler, *Digital Materiality in Architecture* (Baden: Lars Müller; Basel: Birkhäuser, 2008).

51. On the Scholastic-Gothic analogy, as famously suggested by Erwin Panosfky in *Gothic Architecture and Scholasticism*, see above, pages 46–48. With a different bias, similar arguments are suggested by Leach, Turnbull, and Williams in their introduction to *Digital Tectonics*, 4–12; see esp. 4–5 on the new "architect-engineer of the digital age" and on a "certain Gothic spirit" that can be associated with the new digital environment in design and production, "based not on the formal appearance of Gothic . . . but on a certain process-oriented approach towards architectural design, that recognizes the importance of structural force and material composition."

On both the formal and procedural analogies between Gothic and digital, see Spuybroek, *The Architecture of Continuity*, esp. "The Aesthetics of Variation," 245–263, and "Steel and Freedom," 265–284 (and on the rise of a new "Digital Arts and Crafts movement," 283). At the time of writing, Spuybroek is building a "Ruskin Bridge" between Herzogenrath, Germany and Kerkrade, the Netherlands.

52. Lewis Mumford, *Technics and Civilization* (New York: Harcourt, Brace, 1934); see esp. chap. 8, sections 1–2, "The Dissolution of 'The Machine,'" 364–368, and "Toward an Organic Ideology," 368–373.

53. John Ruskin, *The Seven Lamps of Architecture* (London: Smith, Elder and Co., 1849), chap. 5, "The Lamp of Life," XXI, 141: "I said, early in this essay, that hand-work might always be known from machine-work; observing, however, at the same time, that it was possible for men to turn themselves into machines, and to reduce their labor to machine level; but so long as men work *as* men, putting their heart into what they do, and doing their best, it matters not how bad workmen they may be, there will be that in the handling which is above all price ... and the effect of the whole, as compared with the same design cut by a machine or a lifeless hand, will be like that of poetry well read and deeply felt to that of the same verses jangled by rote. There are many to whom the difference is imperceptible; but to those who love poetry it is everything ... and to those who love architecture, the life and accent of the hand are everything." See also Ruskin, *The Stones of Venice*, vol. 3 (London: Smith, Elder and Co., 1853), chap. 4, 194: "The whole mass of the architecture, founded on Greek and Roman models, which we have been in the habit of building for the last three centuries is utterly devoid of all life, virtue, honourableness, or power of doing good. It is base, unnatural, unfruitful, unenjoyable, and impious. Pagan in its origin, proud and unholy in its revival, paralysed in its old age ... ; an architecture invented, as it seems, to make plagiarists of its architects, *slaves of its workmen*, and sybarites of its inhabitants." (Emphasis mine.)

54. Office for Metropolitan Architecture, Rem Koolhaas, and Bruce Mau, *Small, Medium, Large, Extra-Large* (New York: Monacelli Press, 1995).

55. Charles A. Jencks, *The Language of Post-Modern Architecture* (London: Academy Editions; New York: Rizzoli, 1977); Jean-François Lyotard, in *La condition postmoderne* (Paris: éditions de Minuit, 1979), spoke of the "décomposition des grands Récits," or "métarécits" (*La condition postmoderne*, 31). The "end of history" may have been first proclaimed by Jean Baudrillard, *Simulacres et simulations* (Paris: Galilée, 1981), 62–76 (see esp. 70: "l'histoire est notre référentiel perdu, c'est-à-dire notre mythe"). Fredric Jameson's *Postmodernism, or the Cultural Logic of Late Capitalism* (Durham: Duke University Press, 1991; citations are to the 2005 edition), which was for a time influential in some American academic circles, is mostly foreign to the topics under discussion here. Jameson's book is a collection of essays written between 1984 and 1990, inspired, the author claims, by the first architectural definitions of postmodernism (mostly borrowed from Jencks and Robert Venturi), and otherwise unrelated to the philosophical aspects of the matter. Jameson crystallized the stigma still often associated with the early architectural postmodernism of the seventies, as in his view postmodernism stands for little more than a farce of conspicuous consumption, marking the end of the social and aesthetic principles of high modernism, and the simultaneous demise or defeat of socialist ideals. Jameson discusses as icons of postmodern architecture such improbable champions as Gehry's own house in Santa Monica (1979) and John Portman's Westin Bonaventure in Los Angeles (1977). Reference to the latter may have been influenced by Tom Wolfe's unsolicited accolades of the work of John Portman (*From Bauhaus to Our House*, 1981), and Jameson probably saw Gehry's house as a model for a (deconstructivist) strategy of "negativity, opposition, subversion, critique and reflexivity" against the faceless dominance of "late capitalism" (Jameson, *Postmodernism*, 48).

Jameson's rambling, opaque, and self-commiserating tirades bear some responsibility for the Left's persistent unwillingness or incapacity to confront postmodernism in historical and dialectical terms. The two original meanings of "postmodernism," in philosophy and in architectural theory, may have been more closely related than they appear *prima facie*: the first instances of architectural postmodernism in the late seventies were also based upon the demise or rejection of a "master narrative"—the then dominant discourse of architectural rationalism, predicated upon its mandates of technological and social advancement and, in turn, upon an historicist teleology of progress and culmination. However, postmodernist architects famously went on to build or advocate revivalist, premodern architectural and urban forms, whereas postmodern thinkers often strove to interpret or anticipate a new technosocial and economic environment. This original rift was never healed by subsequent architectural criticism, which came to terms with the broader philosophical implications of postmodern thinking but could neither hide nor undo the makings of earlier postmodernist architects. Economists and sociologists have long acknowledged that many of the predictions of postmodern philosophers did indeed come true in the eighties and nineties, and that some philosophical theories of postmodernism underpin—often unsuspectedly—many aspects of contemporary economics, from the financial markets to consumer markets to marketing itself. The economic definitions of "niche markets" and "mass customization," for example, are clear offspring of postmodern theories, and the current (2009) debate on digital culture reenacts in part an older debate on the philosophical categories of postmodernity. The nonstandard, participative turn of today's digital technologies may be shaping the first real postmodern age of architecture ("postmodern" in the sense anticipated by postmodern philosophers).

56. See esp. his *L'illusion de la fin* (Paris: Galilée, 1992).

57. Francis Fukuyama, *The End of History and the Last Man* (London: Penguin Books, 1992). See also his "The End of History?," *The National Interest* 16 (Summer 1989): 3–18.

58. And in architecture, this was no metaphor: some of the technologies that became available at the time were spinoffs from military research projects that had been cut back, and the titanium that was used to bestow a shine on one of the most iconic buildings of the 1990s, the Guggenheim Bilbao, appears to have come at a discount from the military of the former Soviet Union, then strapped for cash.

59. Date of the fall (in fact, the opening) of the Berlin Wall. German *Wiedervereinigung* was officially completed one year later (October 3, 1990).

60. From a December 5, 1996, dinner speech by Alan Greenspan, then chairman of the Board of the Federal Reserve. Greenspan was warning of the risk of a financial overvaluation of what was then known as the New Economy.

61. Greg Lynn's original expression was "Curving away from Deconstructivism." See "Architectural Curvilinearity," in Lynn, *Folds, Bodies and Blobs: Collected Essays* (Brussels: La lettre volée, 1998), 114 (first published in *Folding in Architecture*).

62. See above, section 3.1.

63. The expression is Tom Wolfe's (*A Man in Full*, New York: Farrar, Straus and Giroux, 1998). Oddly, from the opposite end of the political spectrum both Jameson and Fukuyama cite Wolfe as a defining figure of postmodernism (the former), and an interpreter of the rising flood of "megalothymia" that characterizes posthistorical times (the latter). See Jameson, *Postmodernism*, 56; Fukuyama, *The End of History*, 329.

64. Roemer van Torn, "Aesthetik als Form der Politik," *Arch+* 178 (June 2006): 88–93. See also Fredric Jameson, "Marxism and Dualism in Deleuze," in Ian Buchanan et al., eds., *A Deleuzian Century* (Durham: Duke University Press, 1999), 13–36 (originally published as a special issue of *South Atlantic Quarterly*, Summer 1997).

65. See the recent discussions on the so-called "long tail," from Chris Anderson's eponymous article in *Wired* 12, 10 (October 2004), and subsequent book *The Long Tail: Why the Future of Business Is Selling Less of More* (New York: Hyperion, 2006).

66. See Mario Carpo, "The Bubble and the Blob," *Lotus* 138 (2009): 19–26, with further bibliography.

67. See Mario Carpo, "Post-Hype Digital Architecture: From Irrational Exuberance to Irrational Despondency," *Grey Room* 14 (2004): 102–115.

68. See for example Rahim and Jamelle, eds., "Elegance," *Architectural Design* 77, no. 1 (2007).

69. On the origin of this term, see above, chapter 1, note 78.

70. See for example Christopher Hight and Chris Perry, eds., "Collective Intelligence in Design," special issue (AD Profile 183), *Architectural Design* 76, no. 5 (2006); Lucy Bullivant, ed., "4dspace: Interactive Architecture," special issue (AD Profile 173), *Architectural Design* 75, no. 1 (2005); and its follow-up, Lucy Bullivant, ed., "4dsocial: Interactive Design Environments," special issue (AD Profile 188), *Architectural Design* 77, no. 4 (2007). On the "social" use of BIM software, see note 73 below.

71. A development that some have called "The Internet of Things."

72. See the discussion of one of the foundational myths of the Web 2.0, the famous Galton's experiment (1906), in James Surowiecki, *The Wisdom of Crowds: Why the Many Are Smarter Than the Few and How Collective Wisdom Shapes Business, Economies, Societies, and Nations* (New York: Doubleday, 2004), introduction and passim.

73. See Chuck Eastman et al., eds., *BIM Handbook: A Guide to Building Information Modeling for Owners, Managers, Designers, Engineers, and Contractors* (Hoboken, NJ: Wiley, 2008); Phillip Bernstein and Peggy Deamer, eds., *Building (in) the Future, Recasting Labor in Architecture* (New York: Princeton Architectural Press, 2010;

Richard Garber, ed., "Closing the Gap: Information Models in Contemporary Design Practice," special issue, *Architectural Design* 79, no. 2 (2009). Technologists often describe BIM software as little more than a sophisticated file transfer protocol, needed to improve coordination between different technical teams working on the same project—architects, engineers, managers, consultants or contractors—and to facilitate the flow of information throughout the design, building, and delivery process. In this reductive interpretation, BIM software appears as a simple restatement of the digital "file-to-factory" concept, extended from the small scale of prototyping and fabrication to the large scale of integrated project delivery; and the interactive and participatory potentials of these new digital platforms are thus often disregarded. Indeed, in a contrary development, some large offices have recently started to offer comprehensive BIM packages for smaller architectural firms, including software, consulting, and possibly full project delivery (see, e.g., the program of the symposium "Building Fluency: Platforms for BIM Innovation" organized by Gehry Technologies in London in June 2009: online at <gehrytechnologies.com>, accessed July 26, 2009).

74. Unrelated to the vast scholarly literature on the functions and history of architectural models, see Albena Yaneva, "A Building Is a 'Multiverse'," in Bruno Latour and Peter Weibel, eds., *Making Things Public: Atmospheres of Democracy* (Karlsruhe: Zentrum für Kunst und Medientechnologie; Cambridge, MA: MIT Press, 2005), 530–535, and Latour's comment in the introduction: "Who could dream of a better example of hybrid forums than the scale models used by architects all over the world to assemble those able to build them at scale 1?" (ibid., 24).

75. On the pessimists' side, see Andrew Keen, *The Cult of the Amateur* (New York: Doubleday, 2007); Jaron Lanier, "Digital Maoism: The Hazards of the New Online Collectivism," *Edge*, May 30, 2006 (online at <edge.org>, accessed December 23, 2009); for the optimists, see Kevin Kelly, "The New Socialism," *Wired* 17, no. 6 (June 2009): 116–121, and other recent interventions of social media advocate Tim O'Reilly, who is often credited for inventing the term "Web 2.0."

76. On the relationship between the culture of rock music and mechanically reproduced, original sound recording, see María Rosa Menocal, "The Flip Side," in Hans-Ulrich Gumbrecht and Michael Marrinan, eds., *Mapping Benjamin: The Work of Art in the Digital Age* (Stanford: Stanford University Press, 2003), 291–300.

77. Also known as BAT proofs (*bon-à-tirer*, or good to print): Bernard Cerquiglini, *Éloge de la variante: Histoire critique de la philologie* (Paris: Seuil, 1989). See above, page 25.
78. By constrast, print is an asymmetrical, download-only, monodirectional information technology. We can write in our copies of books, of course; but—unlike scribal and digital additions or revisions—these annotations are most likely to remain external and extraneous to the transmission chain: they will not be passed on to other readers.
79. See Eric S. Raymond, *The Cathedral and the Bazaar: Musings on Linux and Open Source by an Accidental Revolutionary* (Beijing: O'Reilly Media, 1999). In Raymond's thesis, however, the bazaar, not the cathedral, is the aptest metaphor for an open-sourced environment. See also Howard Rheingold, *Smart Mobs: The Next Social Revolution* (Cambridge, MA: Perseus Publishing, 2002), esp. chap. 2, "Technologies of Cooperation," 29–62.
80. See Richard Sennett, *The Craftsman* (New Haven: Yale University Press, 2008). Sennett recognizes in the open source movement some of the same communal spirit that used to elevate ancient craftsmanship (23–27), but then goes on to blame digital technologies as the source of most contemporary evils: first among the evildoers, computer-aided design, which, in Sennett's view, prompts architects to forget how to "draw in bricks by hand," hence replacing the architect's hand with hostile machinery and abetting a "disembodied design practice" (41–42).
81. On the affinity between digital technologies and Gothic building, see note 51 above.
82. Bruno Latour, "Why Has Critique Run Out of Steam? From Matters of Fact to Matters of Concern," *Critical Enquiry* 30, no. 2 (2004): 225–248, and "From Realpolitik to Dingpolitik or How to Make Things Public," in Latour and Weibel, eds., *Making Things Public*, 14–43.

CHAPTER 4

1. See Eric S. Raymond, "A Brief History of Hackerdom," in *The Cathedral and the Bazaar: Musings on Linux and Open Source by an Accidental Revolutionary* (Beijing: O'Reilly Media, 1999), 3–18 (first published on the Internet in 1992).
2. Kevin Kelly, "The New Socialism," *Wired* 17, no. 6 (June 2009): 118–121.

3. Mario Carpo, "The Bubble and the Blob," *Lotus International* 138 (July 2009): 19–27 (with further bibliography).
4. See for example Julia Angwin and Geoffrey A. Fowler, "Wikipedia Volunteers Log Off as Web Encyclopedia Ages," *Wall Street Journal*, November 23, 2009, A1–17.
5. See for example the now famed Arduino project and its diverse applications (<http://www.arduino.cc>, accessed December 7, 2009).
6. Thomas Friedman, "The Do-It-Yourself Economy," *New York Times*, December 13, 2009, New York edition, WK9; Cory Doctorow, *Makers* (New York: Tor, 2009); Greg Lynn, Recycled Toy Furniture Installation at the 2008 Venice Biennale (Golden Lion for Best Installation Project). See Greg Lynn's *Form* web site (<http://www.glform.com>, accessed December 7, 2009). However, neither Lynn's use of "reverse modeling" technologies nor Doctorow's fiction involve user-generated content, as Lynn's is an "authorial" design system (albeit based on aleatoric components) and Doctorow imagines that in the future a distributed design and micro-manufacturing process will be exploited for the mass production of standardized gadgets.
7. See Doreen Bernath, "On Architecture of Building the Picture: China and Pictorial Introjection," unpublished PhD thesis, Architectural Association, London, Academic Year 2009–2010.
8. Such as Gehry Technologies: see above, chapter 3, note 73.
9. See for example the section on the .MGX Design Products on the web site of the Belgian firm Materialise (<http://www.materialise.com>, accessed December 7, 2009).
10. Janet Murray, *Hamlet on the Holodeck: The Future of Narrative in Cyberspace* (Cambridge, MA: MIT Press, 1998), 126–154. (First published New York: Free Press, 1997.)
11. At the opposite end of the scale of authorial added value, digital authors that go all the way to design fully parametric, generative notations sometimes choose to limit their production to a closed catalog of "authorized" variations. This approach may make market sense, but it runs counter to the technology of digital making. To go back to the video game analogy: a video game is ideally designed to make other people play, not for the lone entertainment of its designer.

INDEX